Lingaraj Patnaik

Select chapters from the two books
1.
And It Appears Thus..

and 2.
How I Got This Idea?

Non-mathematical and mildly mathematical chapters are included in this selection, employing a minimum of high school geometry and algebra

Dedicated to Jita

Contents

1. Ego and the Three Laws of Motion — 5
2. Excess Intelligence — 11
3. Theory of Ignorance: Part-One: An Estimation of the Quantity of Human Knowledge — 20
4. Theory of Ignorance: Part-Two: The Vicious Circle of Ignorance — 30
5. Physics of the senses — 43
6. WHAT AM I` — 46
7. Physics of the Mind and Consciousness- Part-1 — 49
8. Physics of the Mind and Consciousness- Part-2 — 59
9. Science and Spirituality — 70
10. Mathematics and Spirituality — 76
11. Genesis and Stability of Consciousness: Want-Fields and Skew-Dynamics, Precipitates of Consciousness — 81
12. A Mathematical Basis for Homoeopathy — 91
13. Death Wish of the World Civilization — 99
14. Vertical Pedal Bicycle — 103
15. Transverse Wave Propulsion in Fish and Birds — 106
16. Building Wound-Core Transformer — 115

17. Solution to Nuisance Tripping of Plant Lightning from Fault in Appliances	119
18. Dogma, Laws of Physics and the Baconian Filter	122
19. A Conceptual Error in the Calculation of Length Contraction in Special Relativity	132
20. Sustainable Development, Entropy, Madness and the Atom Bomb	153
21. An Accounting Framework for Energy Conservation	163
22. Tyranny of Errors	169
23. Time Estimation of a Job	173
24. Industrial Organization and Pump–Pipeline Network	178
25. P = Q X R and an Industrial Organization	181
26. A Structural Limitation of Language: One Dimensionality	185
Index	189

EGO AND THE THREE LAWS OF MOTION

The three laws of Newton, which apply to ordinary matter, also appear equally to apply to ego.

Going through certain bad experiences in life I set about abstracting such knowledge and guidance out of such experiences as I could so that the whole of it does not go waste. I found, among other things, that the three laws of Newton, which apply to matter, also apply to our ego equally.

And It Appears Thus...

For reference the familiar three laws of Newtonian mechanics are reproduced below, in simple language:

> First law of Newton: An object either remains at rest or continues to move at a constant velocity, unless it is acted upon by an external force.

> Second law of Newton: The external forces 'F' on an object is equal to the mass 'm' of that object multiplied by the acceleration vector 'a' of the object: 'F = ma'.

> Third law of Newton: When one body exerts a force on a second body, the second body simultaneously exerts a force on the first body equal in magnitude and opposite in direction.

Ego and the Three Laws of Motion

All the three laws of Newton apply to the ego without change: (note, here, *"I"* is the ego, not *"I"*, the self.)

1. The first law of ego: (**Law of will**)
   "I" (the ego) will go on doing precisely what I (the ego) intend to do (including sleeping nicely) in the face of all opposition.

All of the war history of the world is testimony to the validity of this first law of ego as manifested in the careers of war lords and generals.

2. The second law of ego: (**Law of fight**)
   Whoever tries to either push me or oppose me shall have to try really very hard.

Anyone who has attempted change the habit of a recalcitrant child or to change a local work culture in an industry, or established trends in society knows this.

3. The third law of ego (**Law of revenge**):
   Anyone who tries to oppose me shall be taught a lesson he will not forget in a long time.

Law of vendetta. This is a sanctioned way of life with many individuals, families, communities, nations, and people in this world.

It is easy to fill stories for each of the three laws. Please find stories in your real lives, in story books, histories.

We may explore energies associated with ego in line with the three laws as stated above.

Ego and the Three Laws of Motion

Anger and kinetic energy of ego

When an ego is frustrated in its movement, impeded by a barrier, an opposition, the loss of kinetic energy associated with the movement of ego manifests as anger, an acknowledged form of heat energy *(krodhagni)*. This is in keeping with the law of conservation of energy.

Ambition and potential energy

When an ego has an ambition to achieve an object of desire, the ambition is the potential energy that finances its venture.

Conversion of heat energy of anger into useful work

I have repeatedly observed that the energy of anger can be channeled into useful work by patience and skill. It is true, in keeping with the second law of thermodynamics, only a part, say a third of this energy of anger can be transformed into useful work, in the direction of achievement of a mission: the rest shall be partly suffered and partly vented. It is foolish to waste the entire energy of anger made freely available from bad experiences and bad circumstances. It is wise to utilize a good part of this thermodynamic resource as a fuel for doing some useful work, including dismantling undesirable barriers.

> Three laws of ego:
> 1. Law of will,
> 2. Law of fight,
> 3. Law of revenge.

Theory of karma and conservation of energy

Clearly the principle of conservation of energy applicable in the field of matter is same as the theory of karma applicable in the field of consciousness.

There is a peculiarity in the theory of karma: the positive and negative karmas don't cancel out, each has to be reaped separately.

When we recover (reap) energy from a coiled spring, which had been pulled *[say, in the positive direction: Force* displacement $\equiv$ (+F)*(+ds)]* or pushed *[say, in the negative direction, (-)F*(-)ds]* the quantity of energy exchanged during recovery (reaping) is ideally the same and is positive : equal to F*ds .

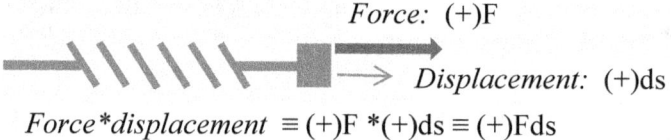

Force: (+)F
Displacement: (+)ds
Force*displacement $\equiv$ (+)F *(+)ds $\equiv$ (+)Fds

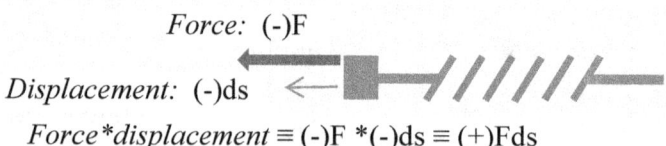

Force: (-)F
Displacement: (-)ds
Force*displacement $\equiv$ (-)F *(-)ds $\equiv$ (+)Fds

During recovery of energy the same quantity of energy exchange has to be suffered as was put into the spring in the first place. Or was put against parts of one's consciousness (in the theory of karma) while as an ego one had intentionally carried out an action. Especially when one acts against the dharma of the consciousness (maya, manifested as the world), the way of its becoming, natural way of unfolding.

Consider that consciousness has its 'natural' dynamics (skew-dynamics), natural way of its becoming, going about its own way, dharma. Thinking as an ego, with the egoistic mind, one may try to push on the consciousness acting with a positive intention; this is like pushing on the spring in the positive direction, (first figure), storing up an energy equal to F*ds , a positive quantity of energy.

Alternatively, going against one's conscience (the inner voice of correct guidance: voice of the existence in an individual which the egoistic mind keeps on smothering by its noise and clamor of ego; this necessitates periodical practice of meditation), one is pushing against one's consciousness in a wrong or negative direction, the energy, thus, stored up (similar to the example of the spring in the second figure) is again a positive quantity, $(-)F*(-)ds \equiv Fds$; it has also to be reaped in time at leisure (separately from the case of action by good intention) subject to skew dynamics involving inertia and capacity.

It is said that the wheels (note wheels, circularity) of karma grind slowly but surely. There is not much escape except in grace, but, that is another matter.

> It is said that the wheels (note wheels, circularity) of karma grind slowly but surely.

Ego, fear and greed and similarity to material force fields of attraction and repulsion

Fear is akin to repulsion and greed is akin to attraction.

It is simple to see the phenomena of fear and greed, i.e., repulsion and attraction respectively applicable to ego as qualitatively similar to the phenomena of attraction and repulsion applicable to mass and charges.

One may, even, see similarities between the distortion caused by ego in the field of consciousness and the distortion of space-time from gravitational mass. Every ego establishes its field of influence around it and can be said to distort the field consciousness thereby and is affected, in turn, therefrom.

EXCESS INTELLIGENCE

In what follows rigor is sacrificed in favor of clarity.

(Will you agree with me that, mathematically.
Δ rigor*Δ clarity ≥ a minimum!)

I always wondered why man is so miserable <u>in spite of</u> being so intelligent and facilitated with so much technology.

I learned that man is so miserable <u>because of</u> being so intelligent and facilitated with so much technology.

I have read somewhere that if a planetary civilization like ours possesses the two technologies of radio telegraphy and atomic bomb simultaneously, then the probability of its survival for the next hundred years is only fifty percent.

And It Appears Thus...

Excess Intelligence Defined

To start with, we define the following three terms:

(i) Acquired intelligence (AI), is acquired by an individual creature by virtue of the evolution of its species;

(ii) Required intelligence (RI), is required for nominal survival of an individual creature;

(iii) Excess intelligence (EI) *is equal to* (=)*the difference between* acquired intelligence (AI) *and* required intelligence (RI) :

$$EI = AI - RI$$

('Nominal' survival? Do not ask for too close a definition. 'Nominal' is what you think is nominal from the context of the discussion and your 'natural' intelligence. Swami Sivananda has said, *"Too much analysis leads to paralysis"*. Just sufficient analysis is to be attempted for the purpose of elucidation of some immediate difficulty that the intellectual, analytical mind erects in the path of an intuitive, holistic understanding. Beyond that point the intellectual, analytic mind is to be discouraged as being profitless).

> *Swami Sivananda has said, "Too much analysis leads to paralysis". Does this chapter qualify?*

Consider a slum dweller. He must make some minimum effort or investment for protecting his meager possessions (horizontal color line). However, as we consider people with greater wealth (inclined color line), the investment to protect a dwelling must increase somewhat proportionately to the wealth. We may graphically plot this requirement as follows:

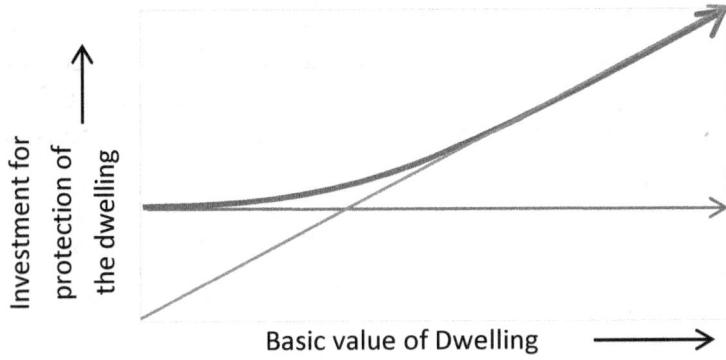

Excess Intelligence

The same principle applies to creatures, small and big with regard to required intelligence: some minimum intelligence is required at the lower end of the scale, say, for example, for the insects for nominal individual survival. Much higher level of intelligence is required, however, for the survival of an individual mammal.

Nature achieves this purpose generally by adding progressively more brain matter as the body weight of a creature increases.

We may, thus, consider the AI, the acquired intelligence, as a function of the number of interactions possible amongst the brain cells:

Number of cells	Number of possible interactions
A:	No interaction
A, B:	(AB), (BA), <u>two</u> interactions
A, B, C:	(AB), (BA), (AC), (CA), (BC), (CB), (ABC), (BCA), (CAB), (BAC), (ACB), (CBA), etc., <u>twelve</u> interactions.

Excess Intelligence

As the brain cells increase in number in a creature, the number of possible interactions (intelligence) increases very much faster than the number. Plotted as a function of body weight, this would, as per the above consideration, appear as:

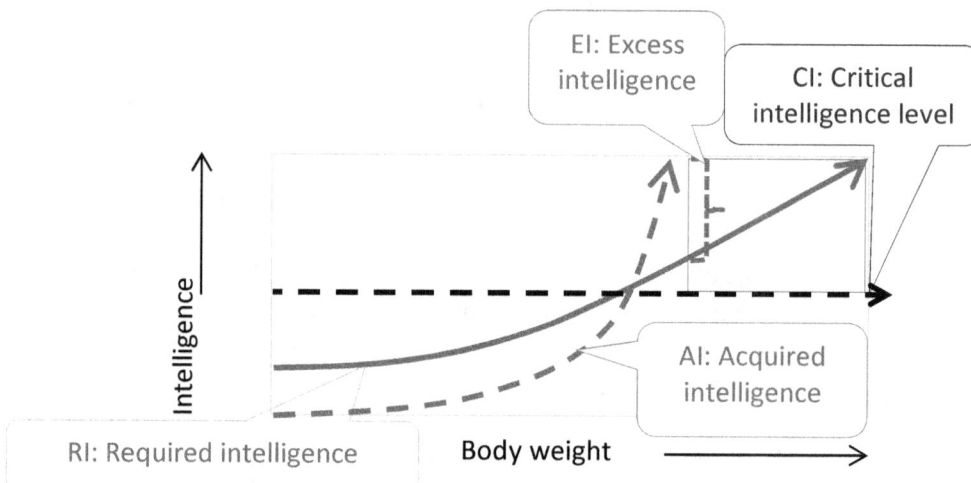

Excess intelligence (EI) *is equal to (=)*
the difference between
acquired intelligence (AI) <u>*and*</u> required intelligence (RI) :

$$EI = AI - RI$$

[There is an interesting associated phenomenon: "Out of Step Evolution of Body vs. Brain". We are, however, not pursuing this matter here.]

Excess Intelligence

The point of intersection between the two lines: one, of required intelligence and two, of acquired intelligence is the point of critical intelligence (CI) where AI = RI , i.e., where the acquired intelligence (acquired by virtue of evolution) is equal to the required intelligence (required for nominal survival of the individual). The reference line drawn horizontally through this point may be called the critical intelligence (CI) level. We shall avoid writing the brackets now onwards.

When $AI < RI$ (AI is less than RI), the survival of an individual creature is precarious, it cannot protect itself from all kinds of environmental factors including other creatures that eat it up. For example, the chance of survival of an individual insect till it dies an uneventful death is very low. It may not possess any sufficient idea of, for example, the birds in its neighborhood which devour it.

> If $AI < RI$:
> Nature ensures survival of a particular species of insects as a whole - in the face of a low probability of survival for the individual - by resorting to a method of procreating in vast numbers.

Nature ensures the survival of a particular species of insects– in the face of a low probability of survival for the individual– by resorting to the method of procreation in vast numbers.

When $AI = CI$, say, for a sizeable creature, like, for example, a monkey or a dog, the creature has a reasonable or 'nominal' chance of survival as an individual.

When <u>AI is greater than CI</u>, the excess intelligence, EI (the difference AI − RI) is positive, and we arrive at the species of *Homo sapiens*, ourselves, at the extreme end of the scale.

Excess Intelligence, the Root Cause of Misery

EI, excess intelligence, is the root-cause of almost all of human suffering. Excess intelligence makes a man overly anxious for the future, suffer futile regrets of the past and prevents a fairly happy living in the present.

Excess intelligence saps one's energy.

As such, the human brain, a mass of tissues without any mechanical movement, representing only 2% of the body weight, consumes twenty percent of the energy used by the human body, more than any other organ (it receives 15% of the cardiac output, 20% of total body oxygen consumption, and 25% of total body glucose utilization). The approximate power consumption of the human brain is 20–40 W. For comparison, the power consumption of a typical household fluorescent tube light is 30–40 W.

> Excess intelligence, is the root-cause of almost all of human suffering. Excess intelligence makes a man overly anxious for the future, suffer futile regrets of the past and prevents a fairly happy living in the present.

Excess Intelligence

The excess intelligence represents an imbalance– especially of an excessive variety *Homo sapiens* are concerned– look at what has happened to the natural environment in six decades flat!

In this context it is interesting to read about the concept of Species Encephalization Quotient (apart from the concept of Brain to Body Weight Ratio) which attempts to describe animal intelligence:

Species Encephalization Quotient

Man	**7.44**
Dolphin	**5.31**
Chimpanzee	2.49
Rhesus Monkey	2.09
Elephant	1.87
Whale	1.76
Dog	1.17
Cat	1.00
Horse	0.86
Sheep	0.81
Mouse	0.50
Rat	0.40
Rabbit	0.40

How Indian Culture Saved the Natural Environment over Five Millennia

There may be many different valid viewpoints in this regard. One is, whereas the western culture, motivated by an egoistic, individualistic, analytical, exploitative outlook, developed technologies which have, in sixty years flat, pushed the natural environment to the brink of a significant irreversible damage, the eastern culture, on the other hand, evolving a spiritual, universal and worshipful, celebratory, holistic approach, helped to tame the redoubtable native intelligence and thereby preserved the environment over five thousand years. The practice of meditation and the religion of treating everything as part and parcel of the same godhead has helped to save the natural environment from being destroyed by the intelligent Indians.

> Prolonged religious rituals including meditation, concentration, rituals from early morning onwards, including repetitive chanting, etc., effectively nullified the excess intelligence EI of the Indians. And preserved the sub-continental environment thereby.

Prolonged religious rituals including meditation, concentration, rituals from early morning onwards, including chanting, etc., effectively nullified the excess intelligence EI of the Indians. And preserved the sub-continental environment thereby.

Amongst many different methods the human beings have adopted to curtail excess intelligence, i.e., relax the mind, are: dozing, gossiping, walking in the bazaar, consumption of drugs, drinking, meditation, etc. No doubt, Marx said that religion is the opium of the people.

Excess Intelligence

Many religious cults are known to encourage the use of opium and such other drugs to attain spiritual high. Many people make direct use of spirit for a high.

THEORY OF IGNORANCE: PART-ONE: AN ESTIMATION OF THE QUANTITY OF HUMAN KNOWLEDGE

atha va bahunaitena kim jnatena tavarjuna
vistabhyaham idam krtsnam ekamsena sthito jagat
BG-10:42

(TRANSLATION
But what need is there, Arjuna, for all this detailed knowledge?
With a single fragment of Myself I pervade and support this entire universe.)

In the area of knowledge several things puzzle me. One is that the whole of the knowledge and experience accumulated up to the moment is not adequate to meet the challenges of the next hour. Second is our self-congratulatory position on our collective achievement of 'progress'.

I set about arriving at a crude estimate of the quantity of human knowledge over the ten millennia.

And It Appears Thus...

Less than a ppm (a part per million)

We shall see that an upper limit for the sum total of human knowledge is not more than 'one part per million' (ppm) of the total possible knowledge of the truth of our existence.

Theory of Ignorance: Part-One: An Estimation of the Quantity of Human Knowledge

Consider the fact that our knowledge of matter (inert, dead matter) is quite incomplete: we do not know such fundamentals as what energy, space and time are. Question closely and you shall soon find out. Opinions you may get aplenty but not understanding.

Fundamentals are at the foundation of our experience.

Incidentals, in contrast, are descriptions of assemblies in space and processes in time.

It is interesting to note that, we, indeed, cannot answer the question *"What is?"* with regard to any of our fundamental experiences: *"What is space?"*, *"What is time?"*, *"What is energy?"*: such fundamental questions have not been answered by anybody till date. For that matter, *"what are mass, charge, potential"*, etc., are also not known. Even, such innocuous questions such as *"what, fundamentally are, beauty, honesty, sweetness, bitterness, happiness, sadness"* cannot be answered: all of these are fundamentals of experiences. There is a great deal of tautology (circularity) and interdependence in the definitions of these fundamentals of experiences. See the circularity of definitions (tautology) in the following:

1. *"What is space?"*
   Answer: Space is that which is measured by a calibrated edge.
   But, a calibrated edge has the concept of space built in to it!

2. *"What is time?"*
   Answer: Measured by clock.
   But, clock has the concept of time built in to it!

3. *"What is energy?"*
   Answer: Energy is that with which we do work. But, work involves only transformation of energy from one form to another! (Say, from heat energy in the petrol fuel to gravitational potential energy when a car climbs up a gradient or simply frictional heat when it runs on a level surface!)

4. *"What is mass?"*
   Answer: Mass is a measure of matter.
   Again tautology! *What is matter?*

5. Again, *what is (inertial) mass?*
   Answer: Inertial mass is a measure of 'inertia' defined by the equation
   $$m = F \div a$$
   inertial mass = force ÷ acceleration.

   But, now *what is force?*
   $$F = ma$$
   *force = inertial mass * acceleration.*

   Nice!

6. *What is gravitational mass?*
   Answer: It is defined by the equation:
   $$F = GMm/R^2$$
   gravitational force between two gravitational masses
   =
   product of the gravitational masses
   ÷
   square of the distance separating the centers of the two gravitational masses.

Now *why are the inertial mass and the gravitational mass equal in magnitude? Are they one and the same fundamentally?* Nobody knows.

More interestingly, this equation ($F = GMm/R^2$) does NOT tell us *"What is gravity?"* nor *"Why gravity operates?"* not even *"How does gravity operate?"*. The equation does not offer a mechanism behind gravity.

All that the equation for gravity ($F = GMm/R^2$) tells us are:
1. The 'magnitude' of the force of gravity and
2. The direction of this force: the force is attractive and is along the straight line connecting the two centers of the gravitational masses.

Period.

7. *What is electric charge?* Nobody knows. We, of course, know that there are two types of electric charges, positive and negative, and that similar charges repel each other but opposite charges attract each other with a force given by

$$F = HQq/R^2$$ (where we write H for $1/4\pi\varepsilon_0$)

The status of answers to the questions *"What?, "Why?" and "How?"* are no better than they are for gravity.

8. There is a great deal of complexity when an electric charge interacts with a magnetic field - it

is not so simple as the interaction between two gravitational masses with each other or two electric charges with each other: the so called Lorentz force comes in to picture where the directions of the force on a charge is perpendicular to both (i) the direction of the movement of the charge and (ii) the direction of the magnetic field!! *Why?* Nobody knows.

(As a young boy, I had come across a description of a generator of hydroelectricity in my class four literature book in Odia language. A hydroelectric generator was described thus: water falls on a turbine and makes it rotate; the turbine, in turn, rotates a magnet, mounted on an axle, within a conducting copper coil; this makes electricity come out of the ends of the copper coil. Fortunately I could understand the set up as was described; it is fairly straight forward, except that I did not understand how the electricity comes out of the end of the copper coil. To date I don't understand fundamentally the mechanism of how the electricity is forced out of the ends of the copper coil by a rotating magnetic field. I can, of course, calculate exactly the emf generated and the current that would flow given the circuit impedances; but nothing tells me precisely what transpires between the cutting of magnetic field by the copper conductor and the generation of the voltage; I do not know the mechanism. The theory stands upon too many assumptions stated as fundamental laws. Besides, what appeared as an explanation yesterday appears as a mere description today.)

9. Even the popular electron is not so familiar to the scientists: see the book "Mathematics for Physics and Chemistry" by Moseley Murphy and Henry Margenau:

> *Electrons are waves and particles together: a monstrous idea.*

Contrast this with the following statement from Wikipedia:

> *Electrons also have properties of both particles and waves, and so can collide with other particles and can be diffracted like light.*

It would seem we know next to nothing! How true! Processes involving cyclones, lightning are not fully understood. What is life, consciousness are not known.

"Mathematics for Physics and Chemistry" by Moseley Murphy and Henry Margenau:
"Electrons are waves and particles together: a monstrous idea."

-However-
IS A FISH MOVING IN WATER A MONSTROUS IDEA?
A swimming fish is, indeed, a macroscopic example of an object which is at once a particle and also a wave.

Theory of Ignorance: Part-One: An Estimation of the Quantity of Human Knowledge

On the material plane, we seem to know very little. The fact that we are able to manipulate the world around us successfully hides the fact of our ignorance. (It is like a skilled car driver who may be blissfully unaware of thermodynamics). Why! Many many engineers operating various industrial dryers and heaters are unaware of thermodynamics!

And most of the engineers are blissfully unaware of the damage they are bringing upon directly to the environment and themselves.

Ignorance also affects biologists busy in solving problems in food production by genetic engineering, and doctors inventing all kinds of health solutions. The foolhardiness with which we are manipulating the world around us, the manner in which we indulge in the large scale invasion of the limited resources of the our poor little earth, without really having any great and in-depth understanding of all the factors is a measure of the pervasive nature of our ignorance.

> Sri Aurobindo said "The advocates of action think that by human intellect and energy making an always new rush, everything can be put right; the present state of the world after a development of the intellect and a stupendous output of energy for which there is no historical parallel is a signal proof of the emptiness of the illusion under which they labour."

Theory of Ignorance: Part-One: An Estimation of the Quantity of Human Knowledge

Jesus might have said of the entire mankind today: OH GOD! FORGIVE THEM BECAUSE THEY DO NOT KNOW WHAT THEY ARE DOING! He might have added "TO THEM SELVES!

Will Jesus' blood suffice to expiate for our sins when the entire planet is burning away in the global warming? And two hundred species become extinct every day? For the vandalization we are in the habit of calling 'our' civilization?

karmaṇy akarma yaḥ paśyed akarmaṇi ca karma yaḥ
sa buddhimān manuṣyeṣu sa yuktaḥ kṛtsna-karma-kṛt
Gita 4.18

(TRANSLATION
One who sees inaction in action, and action in inaction, is
intelligent amongst men, and he is in the transcendental
position, although engaged in all sorts of activities.)

A measure of human knowledge

To recapitulate, on the material plane, we know almost nothing fundamentally about what are space, time, energy, gravity, etc,. However giving some value to the enormity of the effort of mankind over the last ten millennia of meditation and scientific effort, let us say, our knowledge of the material plane is, at most 10%, i.e., 10^{-1} portion of the knowledge of the material plane.

If our knowledge of the material plane equals at most only 10^{-1} part, what about our knowledge of the physical body with its instincts evolved over three million years of evolution? Say, at most 10^{-2} part?

Theory of Ignorance: Part-One: An Estimation
of the Quantity of Human Knowledge

Of our sensoria including the brain? Using same logical trend, at most 10^{-3} part?

Of the mind of which we cannot even answer such simple questions as *'What is mind?'*, *'What is its size? Shape? Color? Weight? Location?'* What we may say is the quantity of knowledge we possess of the mind? Continuing in the same logical fashion, at most 10^{-4} part?

> *Of the mind of which we cannot even answer such simple questions as 'What is mind?', 'What is its size? Shape? Color? Weight? Location?'*

Then of the vitality that sustains mental activity, which is called 'prana', 'qi' or élan vital? At best 10^{-5} part? (Think of energy relating to brain with its electrochemical 'signals' forming the basis of brain's functioning and prana relating to mind with its visual, aural and other 'symbols' forming the basis of mind's functioning.)

Then, what is the quantity of knowledge we have of the core entity, the self, the being, the soul that gives meaning to the complex functioning of the mind and the sensoria? (Consider the lie detector apparatus that indicates the existence of a central being which reveals the registration of a spoken lie in the instantaneous response of ECG and dilation of pupils?) No more than 10^{-6} part? <u>One part per million (ppm)?</u>

Around ten thousand years of meditation and two thousand years of development of science and technology in the nearly thirty thousand years of human presence on

Theory of Ignorance: Part-One: An Estimation of the Quantity of Human Knowledge

this planet has yielded no more than one ppm of the total conceivable knowledge of the Reality.

Drunk with this meager, no more than one ppm (part per million) quantity of knowledge of the reality, and driven by a pride of stupid superiority over the entire creation, we are shoring up vast amounts of reaction from our invasive actions on various global processes which we neither understand nor can face when they confront us in their elemental ferocity. Has time not come to stop being driven by an egotistical, differential, analytical intellect setting the mankind against the entire environment, one nation against neighboring nation, one sect against its sister sect, one family against its neighboring family, brother against brother and sister? Has the time not yet come to transcend the intellectual development which threatens to have become sterile, and, instead, seek an intuitive, holistic, worshipful, meditative, symbiotic, synergistic, celebratory meaning of life? And divine the soul of the existence, to come round the full circle and possibly discover that this universal soul is the very self of our being, seeking ever to celebrate our becoming? And make love and not war?

Earth has already paid a heavy and an unacceptable price, a bloody historical price, for the evolution of intelligence which has lead us to the present stage of intellectual development where we must transcend it and enter in to an intuitive consciousness or else.

THEORY OF IGNORANCE: PART-TWO: THE VICIOUS CIRCLE OF IGNORANCE

Na roopamasyeha tathopalabhyate
Naanto na chaadirna cha sampratishthaa;
Ashwatthamenam suviroodhamoolam
Asangashastrena dridhena cchittwaa.
Gita 15:3

(TRANSLATION
Its form is not here perceived as such, neither its end, nor its foundation or resting place; having cut asunder this firm-rooted Peepul-tree (Ficus religiosa) with the strong axe of non-attachment,)

Na tadbhaasayate sooryo na shashaangko na paavakah;
Yadgatwaa na nivartante taddhaama paramam mama.
Gita 15:6

(TRANSLATION
Neither doth the sun illumine there, nor the moon, nor the fire; having gone thither they return not; that is My supreme abode.)

Aashcharyavat pashyati kashchid enam
Aashcharyavad vadati tathaiva chaanyah;
Aashcharyavacchainam anyah shrinoti
Shrutwaapyenam veda na chaiva kashchit.
Gita 2:29

(TRANSLATION
One sees This (the Self) as a wonder; another speaks of It as a wonder; another hears of It as a wonder; yet, having heard, none understands It at all.)

Andham tamah pravishanti ye avidyam upasate
Tato bhuya iva te tamo ya u vidyayam ratah.
Isha upanishad, 9

(TRANSLATION

Theory of Ignorance: Part-Two: The Vicious Circle of Ignorance

They enter in to blinding darkness who worship avidya; in to still greater darkness, as it were, do they enter who delight in vidya!)

Vidyam cavidyam ca yastad vedobhayam saha
Avidyaya mrityum tirtva vidyaya amrutam ashnute.
Isha upanishad, 11

*(TRANSLATION
One who knows both vidya and avidya together, overcomes death through avidya and experiences immortality by means of vidya.)*

No matter how much one thinks one knows or has experienced, one always stands challenged by the events unfolding the next hour. The compulsions of life and circumstances always challenge one's knowledge, preparations and abilities. It would seem a core of reality remains permanently beyond one's conscious access, experience and understanding. Perhaps this is what Osho called the 'unknowable' amongst the triad: the 'known', the 'unknown' (which shall become known in course of time) and the 'unknowable'.

And It Appears Thus...

Five facts of ignorance

Fact One: All that is true is not experienced.
Fact Two: All that is experienced cannot be described.
Fact Three: All that is described cannot be explained.
Fact Four: All that is explained is not understood.
Fact Five: All that is understood is not truth.

ex|peri|ence
[ɪkˈspɪərɪəns, ɛk-]

VERB

encounter or undergo (an event or occurrence):
"the company is experiencing difficulties"

synonyms: undergo · encounter · meet · have experience of ·

feel (an emotion or sensation):
"an opportunity to experience the excitement of New York"

ORIGIN

late Middle English: via Old French from Latin experientia, from experiri 'try'.

Powered by Oxford Dictionaries · © Oxford University Press

Theory of Ignorance: Part-Two: The Vicious Circle of Ignorance

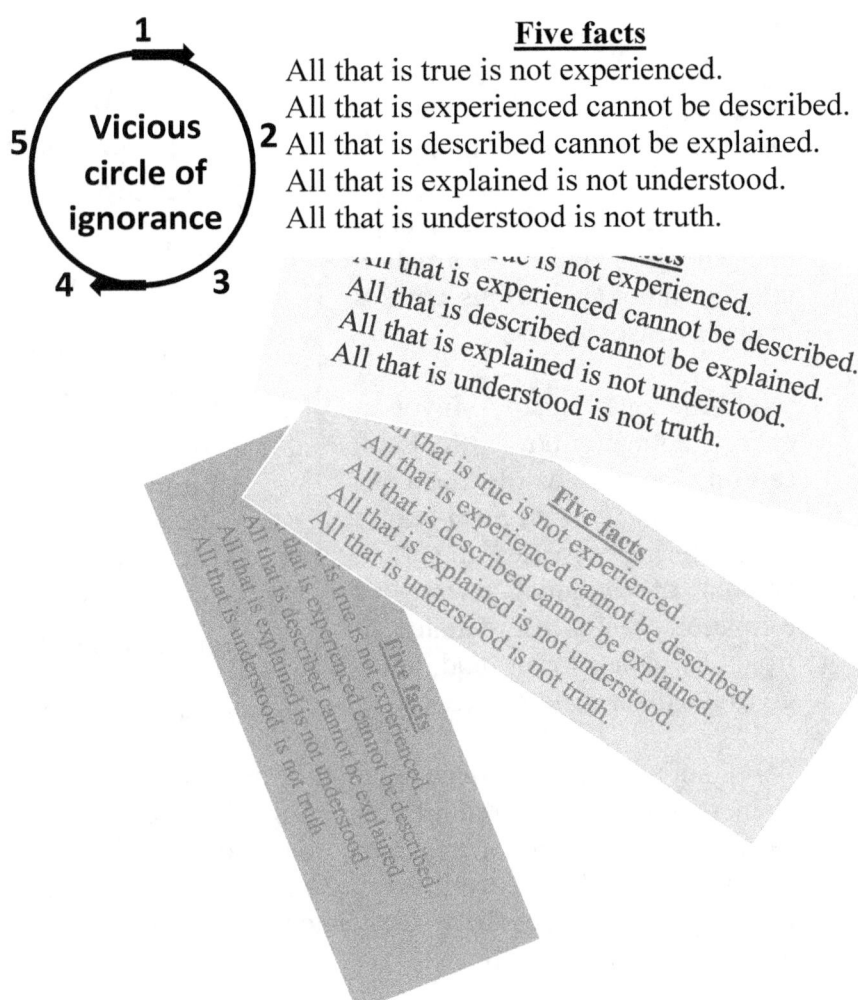

Five facts
All that is true is not experienced.
All that is experienced cannot be described.
All that is described cannot be explained.
All that is explained is not understood.
All that is understood is not truth.

Fact One: All that is true is not experienced

Experience is necessarily a differential phenomenon based on the differentiating principle of ego.

> Fact one: All that is true is not experienced

Experience involves exchange of energy and information together amongst various entities in one's consciousness.

Information also involves energy, no matter how little. Look carefully at the word 'form' in 'in*form*ation': 'in-*form*-ation'. Every form possesses a certain amount of potential energy of formation that differentiates it from the other forms and, thus, gives it a distinctive presence.

> Experience is necessarily a differential phenomenon.
>
> That is why one cannot 'experience' one's own self.

As already said earlier, experience is a differential phenomenon based on the differentiating principle of ego. That is why one cannot experience one's own self. One may be aware of one's selfhood but cannot experience it. Experience is necessarily associated with something different from one's self.

Self can only be experienced asymptotically through the non-self, consciousness, by approaching meditatively, intuitively, silently. A mere analysis will not lead one to experience of the self any more than a mere dissection of her body will reveal one's mother.

Theory of Ignorance: Part-Two: The Vicious Circle of Ignorance

There is nothing truer, finally true, than one's self. This selfhood provides meaning to the variegated experiences, emotions, thoughts, life.

There is no point in asking *'What is Self?'* either. Things which are of a more familiar and common place experience, like space, time, mind, etc. do not yield to such a question. Take for example, *'What is space? time? energy?, mind, beauty? symmetry? honesty? simplicity?'*

All of the above experiences fundamental experiences of our consciousness and cannot as such be defined or explained in simpler terms or analyzed in to simpler components. Any attempt to define these will end up in tautology.

Fundamentals are only to be experienced and accepted as such, not subject to further analysis in terms of other fundamentals.

If a fundamental yields to analysis, can indeed be defined in terms of other fundamentals, it will no more remain a fundamental- it will become a derivative.

> **Fundamentals are only to be experienced and accepted as such, not subject to further analysis in terms of deeper fundamentals; this would render the fundamentals in to derivatives.**
>
> <u>**Analytical mind stops here, intuition takes over.**</u>

<u>Analytical mind stops here, intuitive mind takes over.</u>

Only incidentals, as different from fundamentals, such as assemblies in space or processes in time can be described or defined or explained in terms of component parts. Fundamentals cannot be so defined.

Assemblies in space

What is a vehicle? You can define and explain it as an assembly of parts. You can also define and explain its movements in time.

> *The physical parts of the motorcycle may change, the motorcycle itself may be replaced, but what holds it together is the idea of having a vehicle that will help you save your time, and as long as this idea remains in your mind, you will have a motorcycle that gives shape to the idea. The motorcycle may even evolve into a car, but it's the same original idea that manifests in all these different forms.*
>   *[Jnanagni: The-motorcycle-analogy -to-understand-self-bhagavad-gita]*

Now, let us go back to the 13th verse in chapter 2 of the Gita. The Lord says,

> *The embodied soul, just as it continuously passes from childhood to youth to old age in this body, passes into another body after death. The firm ones never grieve about this.*

While it is not possible to point a material finger in a material world to the self or soul, we may, nevertheless, convince a materialist about the existence of self arguing in the following lines:

1. Yes, we are not able to point out the self or soul. But are we able to point out the commonplace mind, a familiar entity? No. Can we say that the mind does not exist?

2. While we accept that it is the mind that gives meaning to the body with many limbs, we may ask what is it that gives meaning to the mind with its own several functional parts, namely, memory, analysis, synthesis, command, feelings, emotions, instinct, intelligence, intuition, etc.?

Processes in time

Have you cooked a dish? Describe it step by step, this is an example of a processes in time.

Progressively subtler divisions between the world and its observer:

Start dividing the world of experiences thus:

First level of division: World vs. Mind
Second level of division: Consciousness vs. Observer
Third level of division: Awareness vs. Self (Soul)

Notes:
1. In the *Second level of division,* Consciousness means consciousness of the world and mind together: both the world and the mind- together

with the entirety of experiences- are contained within one's consciousness;

2. In the *Third level of division,* Awareness is awareness of the fact mentioned under note 1;

3. In the *Third level of division,* Self or soul is what is called *atman* in Sanskrit, *I* or *me* in English, *me* (*main*) in Hindi, *mu* in Odia, *nenu* in Telugu, *naan* in Tamil. Do not confuse the self with ego.

Fact Two: All that is experienced cannot be described

This is simple. We find that we are in anguish most of the time while we attempt to describe our deeper experiences verbally. We use persuasive body language, smile and make use of other devices to bypass the limitations of verbal language and make it easier to describe our experiences. Much of human drama is centered on this difficulty in correctly describing one's experiences.

> Fact Two: All that is experienced cannot be described

The following is an excerpt from my book *'How I Got This Idea',* last chapter, *'A Structural Limitation of Descriptive Language: One Dimensionality':*

> *The one dimensionality of structure of descriptive language (one can write a book on a thin strip of paper of perhaps ten kilometers long. One word comes after another and is followed by exactly one more word.) makes it difficult to express our experiences which are*

multidimensional. To one who has never seen an aero plane, it is difficult to describe the experience of a flight. However, aided with pictures, the job is easier. It is easiest if one gets hold of a plastic model of an aero plane. However, this is a trivial example. The really beautiful and powerful examples of successful attempts by man at surmounting this limitation of one dimensionality of descriptive language are: poetry, drama, song, dance, mathematics, humor, jokes and the like. For example when one says, "A bird in hand is worth two in bush" one is neither speaking about a bird nor a bush. If one actually tries to describe the idea to another in a plain manner, one will take full five minutes.

To give yet another beautiful example, consider Walter de la Mare's poem "Napoleon" in the same chapter:

> *"What is this world O soldiers! It is I.*
> *I, this incessant snow, this northern sky.*
> *Soldiers!*
> *This solitude through which we go is I."*

This, if we take it purely literally, is pure nonsense. Nevertheless the poet has been able to express high spirituality by overcoming the limitation of one dimensionality of descriptive language.

Even simple sensations cannot be described satisfactorily! Consider describing sweetness of a certain variety of mango to a person.

Or consider describing sex to a four year old.

Indeed, the majority of experiences (sensations, feelings, moods and more) are not describable. Hence, body language, dance, music, mathematics, 3-D (three dimensional) modeling have appeared on the scene.

Fact Three: All that is described cannot be explained

This is very clear. Think of most of your philosophies. Including this one. Think of double talk.

> **Fact Three**: All that is described cannot be explained

Combining the Facts Two *(All that is experienced cannot be described)* and Three *(All that is described cannot be explained)* above, we receive the fact:

All that is experienced cannot be explained.

What appears to be an explanation today in science appears to be a mere description tomorrow.

Indeed, consider the answer a kid provides when asked: *"How the lights go on?"* The kid would perhaps explain by saying *"By putting an electric switch on."* To an engineer, however, this is not a satisfactory explanation. He would explain it in terms of completing an electric circuit containing a power source and a lighting load. To a theoretical physicist even this, the engineer's explanation, may possibly appear as merely a description; he would search for an explanation in a much deeper sense, like, for example, the Maxwell's equations, and possibly deeper as Maxwell's equations may appear as mere description calling for still deeper explanation.

Consider love, beauty. All explanations fail. All descriptions are not equally satisfying. Poetry and drama take over. Language takes on a mystical dimension. Mist is the root word of mystical.

Here it is not out of place to say that, in general, in mathematics, clarity and rigor do not always go together; the more the rigor, the less the clarity: one may even formulate this conceptually:

$$\Delta \text{Clarity} * \Delta \text{precision} \geq \text{a certain psychological minimum.}$$

Fact Four: All that is explained is not understood

This is also very clear. Think again of most of your philosophies. Endless debate ensues because of this fourth fact. Think of the number of papers and books written and the number of papers and books understood. And, misunderstood?

Here is another excerpt from the last chapter of my book *'How I Got This Idea'*:

> **Fact Four**: All that is explained is not understood.

> *All information carrying devices like the radio, the T.V., the computer are subject to NOISE; this is also true about the descriptive language as a medium of transmission of mental imagery. This is one reason why we keep correcting our statements, and*

sometimes entirely scratch and rewrite letters, essays and poems.

All that is understood is not truth

Truth, as we have estimated in an earlier chapter, *'Estimation of the Quantity of Human Knowledge'* is at least a million times more than what we know after ten millennia of meditation and two millennia of science.

If you think you understand a great deal, clearly you are far away from truth. People who know a great deal truthfully admit they know very little.

> **Fact Five**: All that is understood is not truth.

PHYSICS OF THE SENSES

Sensory experiences, well-coordinated as they are, are quite misleading. It is startling to see that forces we perceive with our senses are actually far different from what exist in nature!

And It Appears Thus...

Forces we perceive are far different from the actual!

Sensory experiences, well-coordinated as they are, are misleading. Forces we perceive are actually far different from what exist in nature.

It is well known that our sensoria have logarithmic responses to the magnitudes of various stimuli they receive. This we will return to later.

Apart from the fact of the logarithmic responses of the senses, it is wonderful to see that, in the detail, the nature of forces we experience through our limbs is substantially different from what actually exists in nature.

Consider a pull on your hand while, for example, exerting a steady pull on a suitable object:

Force perceived vs. Actual forces

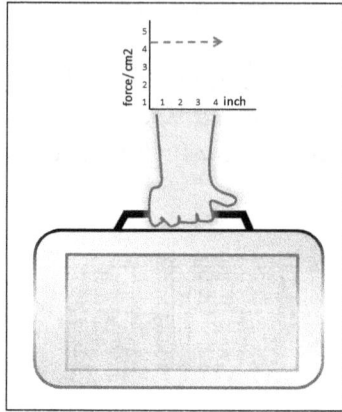 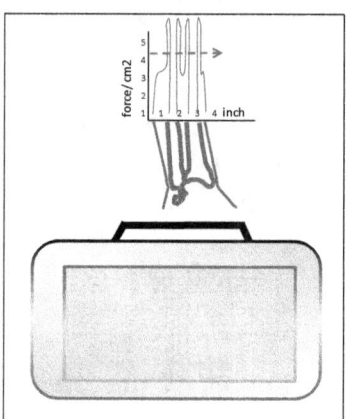

Force perceived
Same quantity of force is 'experienced' at every point on a cross section of the hand

Actual forces
However, the actual force across a cross section of the hand is quite non-uniform: bones and muscles carry the maximum amount of force, skin very little, and bone marrows none.

Logarithmic response of the senses to stimuli

Consider the responses of eyes, ears to brightness or volume of sound – hence the use of the unit of decibel for sound volume. Decibel is a logarithmic unit.

> *The eye senses brightness approximately logarithmically over a moderate range and stellar magnitude is measured on a logarithmic scale.*

Physics of the Senses

Perceived loudness/brightness is proportional to log of actual intensity measured with an accurate nonhuman instrument.
[Wikipedia]

This is a prelude to the next chapter.

WHAT AM I?

Mattah parataram naanyat kinchidasti dhananjaya;
Mayi sarvamidam protam sootre maniganaa iva.
                                      BG 7:7

(TRANSLATION
There is nothing whatsoever higher than Me, O
Arjuna! All this is strung on Me as
clusters of gems on a string.)

There is an easy way to help us realize that, at the bottom, we are not physical entities. All spiritualists from hoary ages have told us this.

And It Appears Thus...

A Caterpillar

> *Caterpillars can eat an enormous amount during a life cycle stage that typically lasts several weeks. <u>Some</u> consume 27,000 times their body weight during this life phase.*
> *[Cool-Facts-About-Caterpillars]*

Consider a baby caterpillar, name it 'Tiny', suitably encouraged to go along a straight line, a mile long: a fresh-n- soft single one-mile-long noodle is laid before the baby caterpillar to feed on in its lifetime.

WHAT AM I?

Caterpillar Tiny on a noodle diet

Cook Zhao Mingxi of One Noodle Restaurant In Chongqing China is reputed to make long single noodles. One noodle of 50 meters in length and 100 grams in weight was made by Zhao Mingxi; he claimed that his record is a 300-meter long noodle.

Let us say, a mile long noodle would be 3.2 kgs. A mile is a little more than 1600 meters. If we assume the body weight of Tiny to be 0.5 gram, then 3.2 kg of noodle would be 6400 times its body weight, which, let us say, is adequate for Tiny's life time.

Let us do an analysis.

After having its bodyweight turned over 6400 times, the worm is still the good old Tiny. At the end of the mile-long-noodle, not even one original atom of its infant body would be found in Tiny.

Just like the grandfather's knife which continues its sacred existence no matter how many times its handle or blade might have been changed time and again.

That should make us think of Tiny as something superior to mere body mass.

WHAT AM I?

At the lowest level, Tiny is NOT a body mass but is a program that upholds a certain resident 'mass' in a body form as Tiny eats along the mile-long-noodle, 6400 times its body weight.

It is same for you and me and the other human beings.

At the lowest level we are not body weights or body masses or plain physical body forms but are programs that operate clutches of material, mental, vital whirlpools we call ourselves..

If we are capable of seeing ourselves as entities beyond the materiality of bodies and body forms, we can go further along the meditative path for self-realization.

We may ask about the entity that is central to the variegated phenomena of thoughts and emotions which we may name as the central entity, the self or soul.

We may go even further and meditate on the basis that provides a stage for all such central entities and their play.

Kavim puraanamanushaasitaaram
Anoraneeyaamsam anusmaredyah;
Sarvasya dhaataaram achintyaroopam
Aadityavarnam tamasah parastaat.
Gita 08:09

(TRANSLATION
Whosoever meditates on the Omniscient, the Ancient, the Ruler, minute than an atom, the supporter of all, of inconceivable form, effulgent like the sun and beyond the darkness of ignorance.)

PHYSICS OF THE MIND AND CONSCIOUSNESS- PART ONE

Chanchalam hi manah krishna pramathi balavad dradham
Tasyaham nigraham manye vayor iva suduskaram
Gita 2:34

(TRANSLATION
The mind verily is, O Krishna, restless, turbulent, strong and obstinate. I deem it as hard to control as the wind.)

It took me more than a decade for the following ideas on mind and consciousness to take shape.

And It Appears Thus...

Working definitions

Mind: Collection of thoughts
Consciousness: Thoughts (mind) + emotions
Maya: Consciousness
Awareness: An uninvolved poise in the consciousness without being clouded by thoughts or emotions

Mind and the flow model: P = Q*R

A human being has some thirty to sixty thousand thoughts in a day. Precisely one thought plays out before and after another thought. Yes, various pressures compete to possess the thought channel and

> A human being has some thirty to sixty thousand thoughts in a day.

propagate their particular thoughts.

This phenomenon can be modeled in many different ways:

Here is an excerpt from my book: *'How I Got This Idea'*, page -127-:

Hydraulics: $\qquad P = Q * R$

*Pressure = Flow rate of water * Resistance in pipe line;*

Electricity: $\qquad V = I * R$

*Voltage = Electric current * Conductor Resistance;*

Thermodynamics: $\quad T = Q * R$

*Temperature = Flow rate of heat * Thermal Resistance.*

> *We notice that, generally, where there is a flow there is a pressure driving the flow against the resistance in the path of the flow. (As an interesting diversion, think of the train of thoughts incessantly going on in your mind, involuntarily. Discover the 'pressure' driving the 'flow of thoughts'. Obviously there is a pressure because the thinking does get accelerated to a high pitch when you are under great pressure. Besides, it takes a great deal of exertion to tame or stem the flow of thought. This may possibly start you on a path of Raja yoga.)*
>
> *One sees that our daily language is a rich reservoir of intuitive models and these can be gainfully harnessed for understanding life. Art precedes, and succeeds science.*

Pressure of work =
*Flow rate of jobs * Resistance to the flow of jobs*

$$P = Q * R \qquad (Law)$$

Indeed, some of the various categories pressures driving the thoughts are pride, lust, fear (three daughters of ego, itself a manifestation of self-doubt). Ego is a wound in selfhood. The current of thoughts in our consciousness are flowing like blood incessantly and ineffectively, driven by pressures (feelings, emotions) of pride, lust and fear, the three daughters of ego, to fill the wound of ego in the self. Grace alone can fill the wound.

Exactly one thought precedes and follows another in the channel carrying the thoughts.

Various pressures (feelings, emotions) constantly compete to take possession of the channel of thoughts.

We may model the mind as a flow of thoughts:

$$P = Q * R$$

where P is the pressure (feelings, emotions), Q is the flow rate of thoughts (i.e., number of thoughts flowing in a minute) and R is the constant of proportionality, the resistance offered to the flow of thoughts.

For me Q is 20 thoughts per minute under normal circumstances. It is at least 3 thoughts per minute sometimes during meditations (excluding the thought that I am observing and counting the thoughts).

We may proceed further. Given a sample human being we may monitor the manner the resistance R varies from hour to hour round the clock, during different phases of the moon, during different seasons, during different stages in one's life (from conception to death), dependence on place, events, experiences, etc.

Taking cue from sciences, we may consider modelling after $I^2 * R$ = power and consider $Q^2 * R$ = a certain quantity of heat power expended in the brain (not mind), entropy, etc.

It is interesting to learn that the brain, the hardware (not the same as the mind, the software) consumes a quarter of the food calories, a mechanically static mass within our skull. And demands a close regulation of temperature and heat dissipation.

> *The brain creates much heat through the countless reactions which occur. Even the process of thought creates heat. The head has a complex system of blood vessels, which keeps the brain from overheating by bringing blood to the thin skin on the head, allowing heat to escape. The effectiveness of these methods is influenced by the character of the climate and the degree to which the individual is acclimatized.*
> *[Wikipedia (Thermoregulation)]*

Here is an excerpt from my book: '*How I Got This Idea*', page -*101*- that relates to entropy:

*[**Entropy and Mind***
I am sorry what I am going to touch upon here may appear out of context. However, I am unable to resist myself. It appears to me that we may be able to apply the concept of thermodynamic entropy conceptually to our mind as well. The idea is this: as we progressively discipline a part of our mind, we throw a correspondingly large

amount of entropy and heat to the rest of it. Look at students furiously preparing for tests. Also think of the common citizen in more developed parts of the world where he has to train himself on ever growing amounts of skills, complexity and coordination of information. This is like walking several tight ropes all at the same time. No wonder, more people are stressed the more a civilization is developed. Perhaps, the concept of Gross Domestic Happiness adopted by Bhutan is superior to GDP.]

Mind as a gas of thoughts

> Chanchalam hi manah krishna pramathi balavad dradham I
> Tasyaham nigraham manye vayor iva suduskaram
> Gita 2:34
>
> *(TRANSLATION*
> The mind verily is, O Krishna, restless, turbulent, strong and obstinate. I deem it as hard to control as the wind.)

Vayu: air, gas.

Let us consider the ideal gas law: $pV = nRT$

Let us see if this equation models the behavior of mind satisfactorily at least in a qualitative way.

<u>Ideal gas law:</u>

$$pV = nRT$$

Here
    P is the pressure of the gas,

V is the volume of the gas,
n is the amount of substance of gas (also known as number of moles),
R is the ideal, or universal, gas constant, equal to the product of the Boltzmann constant and the Avogadro constant,.
T is the temperature of the gas.

Let us say, n is the number of thoughts present in the mind: actually one thought flows at a time in our conscious mind; however a large number of potential thoughts struggle with each other trying to manifest.

If we agree to the idea that mind is a collection of thoughts, then n = the number of thoughts competing to manifest themselves.

Consider R to be an as yet undefined constant in the context of the mind.

Let us test this model qualitatively.

As the pressure p mounts in a crisis, and alternatives thoughts compete within a 'fixed' volume V of mind, we have the following situation:

$p = (nR/V)\ T$ i.e.,

$p = (constant)\ T$ i.e.,

temperature T shall mount proportionally to p.

We know very well that this really happens: brow becomes hot and sweats. This model does predict the trend correctly. However, we do not know how exactly to measure p and V or define them satisfactorily as yet. Some practical clinical measurements may be attempted.

Mind and circular momentum

We are aware of another facet of mind's behavior. As we are attempting to understand the mind, a difficult subject matter, we need not be puzzled if different models we are considering do not appear to jell or harmonize with each other at this stage.

Let us consider the fact that when we struggle with the futility of a repetitive thought, especially when our emotions are hurt, and even when we realize that mind is not offering any new ideas but is merely repeating a thought or a series of thoughts *ad nauseam*, wasting energy, time, attention, clarity; we find we cannot turn off the process except for short periods of time by sheer will.

This is suggestive of an associated rotational inertia of a flywheel:

A great deal of torque is needed to arrest its movement which, instead of damping away, stores up the kinetic energy of rotation elastically as potential energy, to be released in another fresh bout of rotation the moment attention slackens a little. A frictional or damping process is required, instead of a merely reactive or dynamic will, in order to dissipate the kinetic energy of the thought-flywheel of repetitive thoughts. The spiritual discipline of japa, chanting, involving the meditative repetition of a mantra or name of a divine power, is an appropriate spiritual technology here which helps. This is why Krishna says *'Yajnaanaam japayajno'smi'* (repetition of mantra is regarded as the best of all sacrifices) in the Gita 10:25.

Mind and the law of elasticity

Crack or rupture of mind, may be seen to resemble the yield point in the phenomenon of elasticity, which, with in limits, obeys the law

$F = (-)k \Delta x$

which means that a restorative (hence the negative sign) force appears as soon as an elastic object is stretched by an amount Δx ; beyond the yield point, the mind cracks:

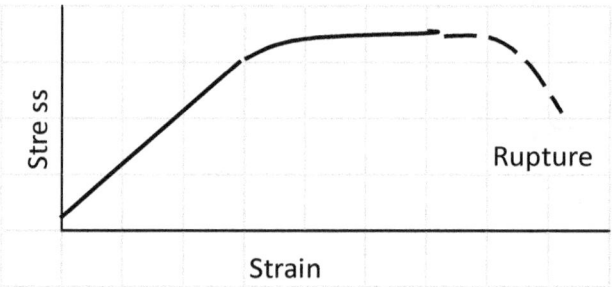

Linear mind, Non-linear life

Whereas senses respond logarithmically to various stimuli, (see Wikipedia for sight, etc.) the mind tends to operate by an overall linear logic. But the mind is always facing the non-linearity character of life in the world. This is one reason why the mind keeps on failing from time to time in taming a non-linear life in the living environment. Perhaps, therein lies a component of *ananda,* delight. If, instead of being constrained by the linear mind we had access to all facts all time, including end results of all the plays, what pleasure would be left?

Think of the 'proportionate' part of the PID control system which, in order to function, requires 'error signal' as the input and responds linearly to this error signal.

Mind modelled after cyclone

> isvarah sarva-bhutanam hrd-dese 'rjuna tisthati
> bhramayan sarva-bhutani yantrarudhani mayaya
> Gita 18:61

(TRANSLATION
The Supreme Lord resides within the hearts of all living entities, O Arjuna, and is orchestrating the movements of all living entities on a carousel of the illusory energy.)

Let us, first, consider the physics expressed in the above verse of the Gita.

A word-for-word translation is as follows:

isvarah	- *the Supreme Lord*
sarva-bhutanam	- *of all living entities*
hrt-dese	- *in the heart*
arjuna	- *O Arjuna*
tisthati	- *resides*
bhramayan	- *causing to move*
sarva-bhutani	- *all living entities*
yantra	- *on a carousel*
arudhani	- *being placed*
mayaya	- *of the illusory energy*

The word 'hrt' is central to this shloka, couplet. To understand this word, consider the sanskrit word *apaharana* which means 'stealing' in English; 'har' or 'hr̃' is the root of this sanskrit word. Same root 'hr̃' also appears to be the root of the word 'hrt' in Sanskrit meaning 'heart'.

Now consider the carousel of the illusory energy maya.

Together, 'hrt' and the carousel of maya are evocative of the image of a cyclone. A depression, a 'stealing' of air mass in the atmosphere, sets the surrounding air in a circular movement which gives rise to a cyclone.

Similarly 'hrt', a stealing, a depression, a want, a desire, a discontentment, an injury (ego) caused to appear in our self together with resources in a circular, repetitive movement of maya, experienced as our consciousness.

The resources of our consciousness also possess inertia of motion, like a flywheel possessing inertia of motion or a cyclonic wind possessing inertia of motion.

[The cyclonic model of the mind, like any other model, physical or intuitive, should not be stretched too far. It is important to remember that every theoretical model is a mental object and, as such, can be stretched or assaulted up to a limit beyond which it will breakdown; that is to say, every mental object, too, has an yield point (static loading) and a fatigue limit dynamic loading) just like any material object. It is common experience that good and sound ideas are, indeed, destroyed every day around us by stretching too far.]

PHYSICS OF THE MIND AND CONSCIOUSNESS- PART TWO

Oordhwamoolam adhahshaakham ashwattham praahuravyayam;
Cchandaamsi yasya parnaani yastam veda sa vedavit.
Gita 15:01

(TRANSLATION
The wise speak of the indestructible peepul tree (Ficus religiosa), having its root above and branches below, whose leaves are the metres or hymns; he who knows it is a knower of the Vedas.)

Yathaa sarvagatam saukshmyaadaakaasham nopalipyate;
Sarvatraavasthito dehe tathaatmaa nopalipyate.
Gita 13:33

(TRANSLATION
As the all-pervading ether is not tainted because of its subtlety, so the Self seated everywhere in the body, is not tainted either.)

Na roopamasyeha tathopalabhyate
Naanto na chaadirna cha sampratishthaa;
Gita 15:03

(TRANSLATION
Its (this banyan tree's) form is not perceived here as such, neither its end, not its origin, nor its foundation, nor its resting place;)

Sarvatah paanipaadam tat sarvato'kshishiromukham;
Sarvatah shrutimalloke sarvamaavritya tishthati.
Gita 13:14

(TRANSLATION
With hands and feet everywhere, with eyes, heads and mouths everywhere, with ears everywhere he exists in the world, enveloping all.)

At Nagercoil, one day, it suddenly dawned upon me that all that I experience is fully contained within my consciousness which is but a portion of me. Not one experience I have is outside of my consciousness. Even the experience of an 'outside world' which is perceived by my mind through the agency of my senses, nervous system and my brain, in its entirety, is but another experience wholly contained within my consciousness.

The experience of an external world perceived by a mind, relating the experience to me, is altogether a misleading experience. The world of infinities is wholly contained within my consciousness. This makes me the all containing illimitable entity. My consciousness, being a portion of me, is relatively a smaller entity providing a platform for the world to play out as a drama, an appearance. As Krishna says, the whole thing stands upside down, 'Oordhwamoolam adhahshaakham ...'

And It Appears Thus...

This and couple of following chapters are the central chapters of this book.

Here is a free play of speculations and plausible explanations.

You are advised to make intuition your guide here. Further, you are advised to adopt an open minded, sympathetic attitude, suspending any critical, analytical approach till after you have read the entire chapter.

Main thrust

Endless experiences accumulate by the hour and tend to misleads one by the energy of their ferocious compulsions. However, all these experiences are entirely, without a single exception, contained within one's own consciousness.

The fact that one's consciousness contains the whole world of experiences is not a statement on the geometrical dimensions of consciousness any more than the geometrical dimensions of a room is a statement on the dimensions of temperature of the room- even though the distribution of temperature is coextensive with the room.

Consciousness may, thus, be coextensive with the material world as experienced without necessarily having geometrical dimensions of the material fundamentally relevant to it.

Consider the following facts (presented as <u>straight forward facts</u>, not assumptions, axioms, hypotheses or postulates).

The sphere of my experience is very well coordinated and compulsive. But it is self-contained in the following sense:

1. There is no single experience whatever which is outside of my consciousness. I carry the entire gamut of my experiences with in the range of my consciousness.

2. I contain my consciousness wholly within me.

3. The experience that there is such a thing as an external material reality which I perceive with my senses is also another experience that is contained within my consciousness.

4. The experience that I have limbs and senses projecting in to an 'outside' world and acting upon it and being acted upon in turn is an experience that is contained wholly within my consciousness.

> Consciousness may, thus, be coextensive with the material world as experienced without necessarily having geometrical dimensions of the material world fundamentally relevant to it.

From the above facts I logically conclude that all experiences, good, bad or indifferent, are contained well with in me. There is no experience outside of me.

Who Perceives?

A dialogue between a sage and a disciple:

> Sage: *I will pose a series of questions before you. There is no trickery involved here. The questions are plain and simple. Please try to answer with simplicity and good faith.*

First question

Sage: *Now, choose between (i) the eyes and (ii) the brain: who or what perceives the sight of brightness, color and texture you experience here and now?*

Disciple, after much effort, replies: *The brain.*

Second question

Sage: *Please take note that the brain is confined inside a hard skull from birth death. No light ever shines upon it (unless it spills out on the road, say, in an unfortunate accident). What reaches the brain is a randomized electrochemical signal from the retina, sent in response to the photons of light falling on the retina. This makes the two things, namely (i) the photons of light which are falling on the retina and*

> The ultimate truth to Nagarjuna is the truth that everything is empty, sunyata, of an underlying essence. Sunyata itself is also "empty," 'the emptiness of emptiness', which means that sunyata itself does not constitute a higher or ultimate 'essence' or 'reality'.
>
> Nagarjuna's view is that "the ultimate truth is that there is no ultimate truth"

(ii) the electrochemical signals which are leaving the retina on their way to the brain two different things.

The brightness, colors and texture perceived are, certainly, related to the electrochemical signals that reach the brain, and may even have one-to-one correspondence. However, the brightness, colors and texture perceived are different from the electrochemical signals.

Considering the details described above, who or what perceives the visual experience?

Disciple <u>replies</u>: *Consciousness, Sir?*

The above discussion between the sage and the disciple is true about every other sensory experience- auditory, olfactory, taste, tactile.

Now, there is no point in asking such questions as "what is consciousness?" The futility of asking 'what' with regard to fundamentals of experience has been dealt with extensively in an earlier chapter, *'An Estimation of the Quantity of Human Knowledge'*.

Taken together, the last two sections add significance to modern physics as follows:

Consciousness-centric Theories of Physics

Consciousness is progressively assuming the center stage in physics.

Physics of the Mind and Consciousness- Part Two

We will treat the observer and the consciousness (which belongs to the observer as a portion) as separate objects.

These two are, however, freely mixed up in literature of contemporary physics, reproduced below from mainly the Wikipedia, where we will mentally substitute the word 'consciousness' in place of the word 'observer'.

Relativity:

> *The second postulate of special relativity:*
> The speed of light in a vacuum is the same for all observers, regardless of the motion of the light source.
>
> [Wikipedia]
>
> *Observer (special relativity)*
> In special relativity, an observer is a frame of reference from which a set of objects or events are being measured. Usually this is an inertial reference frame or "inertial observer". Less often an observer may be an arbitrary non-inertial reference frame such as a Rindler frame which may be called an "accelerating observer".
>
> Einstein made frequent use of the word "observer" in his original 1905 paper on special relativity and in his early popular exposition of the subject. However he used the term in its vernacular sense, referring for example to "the man at the railway-carriage window" or "observers who take the railway train as their reference-body" or "an observer inside who is equipped with apparatus". Here the reference body or coordinate system—a physical arrangement of meter-sticks and clocks which covers the region of space-time where the events take place—is distinguished

from the observer—an experimenter who assigns space-time coordinates to events far from self by observing (literally seeing) coincidences between those events and local features of the reference body.

[Wikipedia]

Quantum Mechanics:

(i) *A father of modern physics, Dirac, says, in the preface to first edition of The Principles of Quantum Mechanics: "Further progress lies in the direction of making our equationsinvariant under wider and still wider transformations. This state of affairs is very satisfactory from a philosophical point of view, as implying an increasing recognition of the part played by the observer introducing regularities that appear in his observatios, and a lack of arbitrarynes in the ways of nature."*

(ii) According to the (Copenhagen) interpretation, the interaction of an observer or apparatus that is external to the quantum system is the cause of wave-function collapse, thus according to Paul Davies, "reality is in the observations, not in the electron".

[Wikipedia]

(i) "The 'reduction of the wave-packet' does take place in the consciousness of the observer, not because of any unique physical process which takes place there, but only because the state is a construct of the observer and not an objective property of the physical system" Hartle, J. B.

Physics of the Mind and Consciousness- Part Two

(1968). Quantum mechanics of individual systems. Am. J. Phys., 36(8):704– 712.

[Wikipedia]

<u>What counts as an observer in quantum physics?</u>

*Observer is a special person (or a system that contains such person) which **does not obey the usual laws of quantum mechanics**. While it is much easier to define observer from a philosophical point of view, the mathematical answer is that the observer is a system which manifests subjective decoherence when observed.*

Critics of the special role of the observer also point out that observers can themselves be observed, leading to paradoxes such as that of Wigner's friend; and that it is not clear how much consciousness is required ("Was the wave function waiting to jump for thousands of millions of years until a single-celled living creature appeared? Or did it have to wait a little longer for some highly qualified measurer - with a PhD?").

[Google]

<u>Observer effect</u>

In science, the term observer effect refers to changes that the act of observation will make on a phenomenon being observed. This is often the result of instruments that, by necessity, alter the state of what they measure in some manner. A commonplace example is checking the pressure in an automobile tire; this is difficult to do without letting out some of the air, thus changing the pressure.

[Wikipedia]

Electromagnetism

The observer who is at rest with respect to a magnet will not see any electric field, but all the other observers moving relative to the same magnet at different velocities shall see different mixtures of electric and magnetic fields.

Classical mechanics

If a solid object is thrown from an aero plane, it will appear to follow a parabolic trajectory to a person standing on the ground, but to a passenger aboard the aero plane, the trajectory shall appear as a straight line pointing downward:

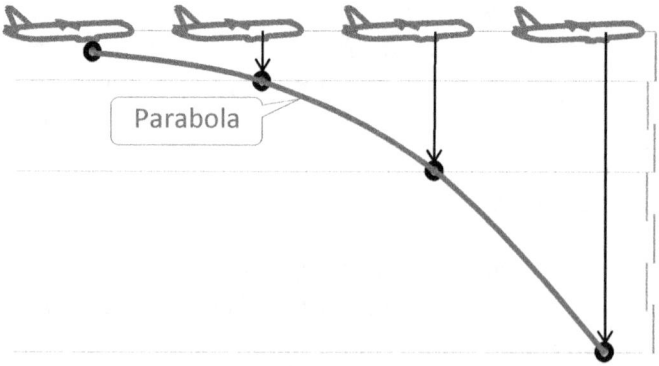

Look at the following most fascinating copy of a letter to the editors of Scientific American:

"Jaron Lanier's description of the seven-camera tele-immersion project in "Virtually There" [April] should have mentioned, for historical context, the traditional two-camera system that has a 20-millisecond latency: the system whose two cameras are called eyes and that uses a computer called a brain on which runs the ever popular Mind OS software that portrays external reality as a near-real-time, three-dimensional, internal representation viewed by ... the mysterious viewer called consciousness."

The editors reply play fully:

"Okay Robert Burruss of Chevy Chase, Md., consider it mentioned. For discussions of other topics from the April issue, please direct your OS below."

SCIENCE AND SPIRITUALITY

Shraddhaavaan labhate jnaanam tatparah samyatendriyah;
Jnaanam labdhvaa paraam shaantim achirenaadhigacchati.
Gita 04:39

(TRANSLATION
The man who is full of faith, who is devoted to it, and who has subdued all the senses, obtains (this) knowledge; and, having obtained the knowledge, he goes at once to the supreme peace)

Bahoonaam janmanaamante jnaanavaanmaam prapadyate;
Vaasudevah sarvamiti sa mahaatmaa sudurlabhah.
Gita 07:19

(TRANSLATION
At the end of many births the wise man comes to Me, realizing that all this is Vasudeva (the innermost Self); such a great soul is very hard to find.

[A video in the same name has been uploaded in Youtube; *Lingaraj Patnaik, Science and Spirituality*]

Many times people inappropriately compare science with religion. Science is a search for truth. Religion is a way of living, to achieve certain specified individual and social goals. Religion is akin to engineering, a way of doing things, method, a plan, a way to be engineered amongst many different available ways to achieve certain specified engineering goals.

Further thoughts led to the following chapter.

And It Appears Thus...

Basics

Science and Spirituality

Science is to be compared with spirituality and not with religion.

The aim of both science and spirituality is to discover the final verities, facts relating to truth, the basic reality of existence as well as the manifestation of the world of experiences.

Religion and engineering are comparable: both are arts- arts of living and arts of doing things respectively.

Wherever methods are employed, even in science, for example, laboratory methods of experimentation, there is art.

In actual practice science and art commingle. Every scientific <u>theory</u> is an invention of the human mind. And, like any other invention, is subject to further improvement as time passes.

Contrast this with <u>facts</u> of science or spirituality which are eternal facts and which do not require further improvement, they are only required to be discovered (uncovered) and understood.

We speak of inventions improving he state of the art.

Religion is to be compared with engineering.

> **Science is to be compared with spirituality and not with religion.**
>
> **Religion and engineering are comparable: both are arts- arts of living and arts of doing things respectively.**

Technology should be compared with specific religious sacrifices like yajnas, homas which are expected to yield specific results. Religion makes use of all such practices just like engineering makes use of specific technology (packages) as may be available to achieve specific goals.

Truth is one, truth is not two: divided in to scientific truth and spiritual truth. There may be two different approaches to truth, one scientific and the other spiritual, but truth is one and the same whether we approach it through the physical sciences of the manifested of nature or explore through the spiritual sciences which have a vaster scope of the existence of which nature is one manifestation, consciousness is another manifestation. Whereas physical sciences deal with nature, spiritual sciences deal with existence as a whole.

> **Truth is one, truth is not two: divided in to scientific truth and spiritual truth.**

Both spirituality and science stand on pillars of faith.

Spirituality stands on the faith that there is existence, you exist, and the fact that you are conscious of your existence.

> **Both spirituality and science stand on pillars of faith.**

And science, especially physics, which is the most fundamental of all sciences, is based on statements of faith which are also known as the fundamental laws of nature.

For example, consider the three laws of Newton, namely (i) the law of inertia, which states that a body keeps moving in a straight line unless and until it is disturbed by an external force, (ii) the second law, which states that force is proportional mass times acceleration, or (iii) the third law, which states that action and reaction are equal and opposite to each other or (iv) the law of gravity which states that the force between two masses is equal to the product of the masses and inversely proportional to the square of the distance between the centers of the two masses, or for that matter, the laws of thermodynamics which comprise (v) the law of conservation of energy apart from (vi) the law of partial conversion of heat in to work and the connected concept of entropy or disorder which monotonously increases in any isolated adiabatic system, or, the two laws of special relativity which are (vii) all the laws of physics are the same in all inertial frames of reference, and (viii) the velocity of light is constant, nearly three lac kilometers per second, for every observer in every inertial frame, or the law of general relativity (ix) that there is no difference gravitation and acceleration of inertial frames in infinitesimally small space-times, or for that matter, the laws of quantum mechanics, which state that (x) all the values of observables measured are the eigenvalues of the corresponding operators. We may also add Maxwell's equations, Schrödinger's equations, Heisenberg's uncertainty principle, etc.; NONE OF THESE LAWS STATED ABOVE HAS A PROOF. All of these are motivated by the experiences of scientists on their observations of nature. Their intuitive minds grasp these verities and these laws are stated and falsifiable predictions are verified and if found correct, the laws are used as pillars of faith on which the rest of the science stands as a superstructure, and is constantly developed.

A successful experimental verification of a falsifiable prediction does NOT constitute a proof; a billion successful experiments also do not constitute a proof: who knows! The next one experiment may fail!

In spirituality the method is the same.

The seers intuitively perceive the verities which they state and then set about engineering ways of life which are called religions for the people at large to follow.

Hence religions differ from place to place and time to time but the science of spirituality is the same every time everywhere.

Now see the methods in science and spirituality:

Scientists employ experiments in the manifested nature. The seers use their experiences in their consciousness and classify, analyze, categorize the results and search for still deeper facts and thus develop the science of spirituality.

> A successful experiment is NOT a proof, a million successful experiments also do not constitute a proof: who knows! The next experiment may fail!

> Thus we see now that both science and spirituality stand on faith as pillars and are developed either by experiments in the area of physical sciences or experiences in consciousness in the area of spirituality.

Thus we see now that both science and spirituality stand on faith as pillars and are developed either by experiments in the area of physical sciences or experiences in consciousness in the area of spirituality. And both attempt to attain to the same final truth, final verity, reality through the methods of science in manifested nature or methods of spirituality in the field of existence.

Note, once a law can be derived from fundamental laws, it is no more a fundamental law, now it is a derived law. For example, the ideal gas laws are derived laws (derived from Newton's laws of motion in statistical mechanics).

MATHEMATICS AND SPIRITUALITY

Mahaabhootaanyahankaaro buddhiravyaktameva cha;
Indriyaani dashaikam cha pancha chendriyagocharaah.
Icchaa dweshah sukham duhkham sanghaatashchetanaa dhritih;
Etat kshetram samaasena savikaaramudaahritam.
Gita 13:06, 07

(TRANSLATION
Earth, water, fire, air, ether; egoism, intellect and the unmanifested
Nature; the ten senses: eyes, ears, tongue, nose, skin and the five
organs of action: hand, feet, mouth, anus, generative organ; the
mind; color, sound, taste and smell touch; desire, hatred, pleasure,
pain, the aggregate body, fortitude and intelligence
—the field has thus been described briefly with its modifications.)

Sattwam rajastama iti gunaah prakriti sambhavaah;
Nibadhnanti mahaabaaho dehe dehinam avyayam.
Gita 14:05

(TRANSLATION
Purity, passion inertia—these qualities, born of nature,
bind fast in the body, the embodied, the indestructible!)

Rajastamashchaabhibhooya sattwam bhavati bhaarata;
Rajah sattwam tamashchaiva tamah sattwam rajastathaa.
Gita 14:10.

(TRANSLATION
They are not constant. Sometimes Sattwa predominates and
at other times Rajas or Tamas predominates. One should
analyse and stand as a witness of these three qualities.)

As different from the material world, where quantity rules (with associated units of measure), in the field of consciousness, quality rules. We must keep this difference in mind while we discuss the topic of mathematics and spirituality.

We cannot, however, get away from the fact that the so called objective material world is but an experience, no matter how compulsive, in the field of subjective consciousness. All the concepts we habitually apply in the objective, material world have their origin in the field of subjective consciousness: measurement, science, philosophy, everything. The subjective experience of a unit of measure, say a meter scale, employed in the world of matter is not fundamentally different from the subjective experience of a meter scale employed in a dream.

And It Appears Thus...

An important question to consider is what kinds of mathematics is applicable in the field of spirituality. Can we apply all of the mathematics we habitually apply in the field of matter?

We may list here various considerations like
1. enumeration,
2. quantitative measurement and
3. qualitative experiences:

1. Enumeration
   Counting various elements is part of the traditional Samkhya philosophy:

   > *Sāmkhya philosophy regards the universe as consisting of two realities; puruṣa (consciousness) and prakṛti (matter). Jiva*

(a living being) is that state in which puruṣa is bonded to prakṛti in some form. fusion, state the Samkhya scholars, led to the emergence of buddhi ("intellect") and ahaṅkāra (ego consciousness). The universe is described by this school as one created by purusa-prakṛti entities infused with various permutations and combinations of variously enumerated elements, senses, feelings, activity and mind. During the state of imbalance, one of more constituents overwhelm the others, creating a form of bondage, particularly of the mind. The end of this imbalance, bondage is called liberation, or kaivalya, by the Samkhya school.

[Wikipedia]

2. <u>Quantitave measurement</u>

We may logically consider the material word experience as a subset of the experiences in the field of consciousness.

Considered thus, any act of measurement in terms of units of measurement, say meters, kgs or seconds amounts to be subjective experiences in the consciousness.

Consider a carpenter measuring out the dimensions of a window in his dream. The unit of measurement represented by his measuring tape in his dream is as real or unreal an experience as one in his wakeful experience of engaging in a similar activity.

We may, thus, consider that, ultimately referred to the gamut of experiences of the consciousness, all experiences are basically qualitative, guna related. Quantitative measurements if any are relative with in the material experiences which, as a whole, are a subset of qualitative, ultimately subjective experiences.

Going further, a scientist or a mathematician is quite justified if he claims that science and mathematics as he receives are subjective experiences and as such must satisfy his subjective sensibilities.

3. <u>Qualitative experiences: Gunas or quality as contrasted quantity</u>

The following excerpt is from an earlier chapter: Theory of Ignorance: Part-Two: The Vicious Circle of Ignorance:

> *Experience is necessarily a differential phenomenon based on the differentiating principle of ego.*
>
> *Experience involves exchange of energy and information together amongst various entities in one's consciousness.*

Qualitative experiences, experiences being differential phenomena, lend themselves to both enumeration and comparison. But qualitative experiences basically do not lend themselves to measurements against any units of measurements.

Quantitative calculations in the field of consciousness

We say one dish is tastier than the other. We may even assign marks to different dishes in a cooking competition. But the marking process is a subjective process, in a field of qualities, not an objective measurement against a standard unit of measurement.

> We will do well to remember always that the so called material world is but a subjective experience of our consciousness.

As long as we accept the subjective process of assigning marks to grade quality, we may attempt to, with some degree of trepidation, use quantitative mathematics in the field of consciousness as we use mathematics in the material world. We will do well to remember always that the so called material world is but a subjective experience of our consciousness.

GENESIS AND STABILITY OF CONSCIOUSNESS: WANT-FIELDS AND SKEW-DYNAMICS, PRECIPITATES OF CONSCIOUSNESS

Eeshwarah sarvabhootaanaam hriddeshe'rjuna tishthati;
Bhraamayan sarvabhootaani yantraaroodhaani maayayaa.
Gita 18:61

(TRANSLATION
The Lord dwells in the hearts of all beings, O Arjuna,
causing all beings, by His illusive power, to revolve as if
mounted on a carousel!)

Mama yonirmahadbrahma tasmin garbham dadhaamyaham;
Sambhavah sarvabhootaanaam tato bhavati bhaarata.
Gita 15:03

(TRANSLATION
My womb is the great Brahma; in that I place the germ [of
desire]; thence, O Arjuna, is the birth of all beings!)

For a long time I had felt that all fields in nature are 'want' fields in the sense that gravitation 'wants' matter, sun 'wants' earth, earth 'wants' the moon and all the other objects. In this sense, the energy associated with a field is a 'want-energy'. Gravitation, Electric and magnetic fields are all 'want-fields'.

Associated the 'want-field' concept is the 'skew-dynamics' concept which, owing to inertia, capacity and such other principles described in this chapter, prevents a straight forward fulfilment of the want-fields. For example, because of skew-dynamics, involving inertia, the earth does not fall into the sun.
This way of thinking has an advantage in describing physics of matter and consciousness and their stability.

Mind, consciousness, maya are treated here in this chapter as synonymous.

And It Appears Thus…

Gravitational field

We may think of a gravitational field as a 'want-field' always seeking all the matter in the universe to fill it. The energy associated with a gravitational field, a positive quantity, is associated with this 'want-field'. Think of a stretched rubber sheet:

We may think of a want-field as a stretched rubber sheet. Each part of the sheet wants all the rest of the sheet to come together.

Electrostatic field

Whereas all matter attract each other with gravity, all charges do not attract each other: like charges repel. To represent such want-fields, think of a large rubber sheet, stretched and clamped just tight, with a special kind of glue sprayed on it. The glue has a peculiar property: each drop is of the same size and spreads on a certain area on the sheet, and as it dries up, shrinks to zero size along with a portion of the sheet in its grasp:

Genesis and Stability of Consciousness

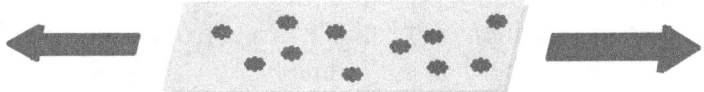

This causes relative stretching of the sheet.

Now consider pulling any two dried glue drops together: they will appear to *repel* each other.

Dissimilar charges' attracting each other, like matter in a gravitational field, is another example of want-field.

Skew-dynamics

Skew-dynamics may be said to manifest when an object sought by a want-field (potential well) is deflected from fulfilling the want-field leading to eventual collapse. Skew-dynamics lends stability to consciousness in the face of want-fields in the consciousness. We may refer to the factors, of which there are several, preventing collapse of the systems as '*barriers*' including, '*skew-dynamic barriers*'.

<u>Solar system</u>
[force is the want-field (potential well) and the inertia of mass of the earth is the skew-dynamic barrier to collapse of the earth orbit]

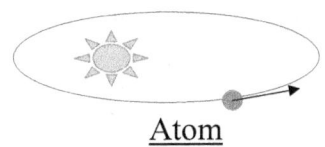

<u>Atom</u>

[The uncertainty principle may be seen the skew-dynamic barrier which prevents electrons from collapsing in to nuclei]

$$\Delta x\, \Delta p \geq \hbar/2$$

(This was introduced in 1927, by the German physicist Werner Heisenberg. It states that the more precisely the position of some particle is determined, the less precisely its momentum can be known, and vice versa.)

1st Orbital of Atom

[Pauli's exclusion principle may be seen as a the skew-dynamic barrier that prevents similar electrons from occupying the same orbital together]

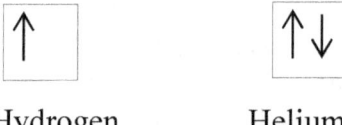

Hydrogen   Helium

(The Pauli Exclusion Principle states that, in an atom or molecule, no two electrons can have the same four electronic quantum numbers. As an orbital can contain a maximum of only two electrons, the two electrons must have opposing spins. This means if one is assigned an up-spin (+1/2), the other must be down-spin (-1/2)).

Varieties of the barriers: Potential barriers (wells), kinetic barriers (skew-dynamic), thermodynamic barriers (skew-dynamic) lending stability to consciousness

Genesis and Stability of Consciousness

Let us consider the potential and the skew-dynamic barriers that prevent collapse of the consciousness back in to the self.

An important question to ask is why the consciousness does not collapse back in to the self. Where is the clue to its mysterious stability in time (time is one experience with in the consciousness)?

Once a consciousness arises out of the self there should be compelling reasons for its continuity in time: its stability. (Here we are concerned with the stability of the consciousness as a whole irrespective of the transient, unstable, processes that it may harbor). Otherwise a consciousness, soon after it arises, shall collapse back in to the self. Quite likely, myriads of consciousness do emerge and collapse back in to the self all the time. One survives.

> Once a consciousness arises out of the self there should be compelling reasons for its continuity in time: its stability.

> An emergent consciousness must, as a whole, be locked inside a potential barrier (potential well).

An emergent consciousness must, as a whole, be locked inside a potential barrier, potential well. Something like mechanical locks which provide *potential barriers* in the form of springs to prevent levers from getting free.

We may, further, conceive of kinetic barriers, or, rather, kinetic skew-dynamic barriers: inertia of earth revolving around the sun provides a *kinetic skew-dynamic barrier*:

this, based on the law of inertia, the first law of Newton, prevents collapse of the orbit of earth.

Apart from kinetic (skew-dynamic) and potential barriers (potential wells) we may also conceive of the thermodynamic (skew-dynamic) barrier. This is, however, partly a variety of potential barrier, associated with the concept of irreversibility in thermodynamics and partly a skew-dynamic kinetic barrier associated with the inertia of the 'particles':

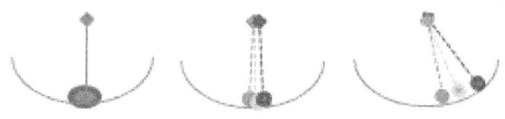

Once pulled to one side and released, as in the last figure, a loose bunch of weights shall, after a few oscillations, start colliding with one another, get randomized i.e., thermalized and are, unlikely to climb back to the top position, together, as the original bunch of weights in the first of the three figures. This kind of phenomena can lock up energy thermally, i.e., in the randomness of movement, and prevent a complete recovery of energy, thus constitute a thermal (skew-dynamic) barrier to the collapse of consciousness back in to the self.

Time is a manifestation of endlessly futile play of restorative processes of skew-dynamics in the want-fields in consciousness.

Stability of consciousness

Though consciousness harbors endless unstable processes, consciousness, as a whole, has continuity, therefore, stability.

This is owing to the genesis of consciousness in a primal want (potential well). A want-field is inherently stable owing to the principle of negative feedback.

If the genesis of the consciousness were not fundamentally with a want-field (potential well) but with, say, a fundamentally an 'explosive field' associated with, say, a 'potential-hill' instead, it would be unstable for lack of negative feedback.

> *Negative feedback occurs when some function of the output of a system, process, or mechanism is fed back in a manner that tends to reduce the fluctuations in the output, whether caused by changes in the input or by other disturbance.*

> *Whereas positive feedback tends to lead to instability via exponential growth (*see 'explosive field' stated above*), oscillation or chaotic behavior, negative feedback generally promotes stability. Negative feedback tends to promote a settling to equilibrium, and reduces the effects of perturbations. Negative feedback loops in which just the right amount of correction is applied with optimum timing can be very stable, accurate, and responsive.*

> *Negative feedback is widely used in mechanical and electronic engineering, but it also occurs*

naturally within living organisms, and can be seen in many other fields from chemistry and economics to physical systems such as the climate. General negative feedback systems are studied in control systems engineering).

[Wikipedia]

Entire manifestation, inclusive of explosive phenomena is contained in an overall want-field, potential well, of one's consciousness.

A plausible mechanism for the genesis of consciousness, maya in terms of the conceptual rigidity of the number '1' (one)

Between physical matter and physical space, both of which are objects of the consciousness, the latter, namely the physical space is perhaps more rigid. Consider the following two facts of nature: (a) physical space harbors the highest speed of wave (light), namely 300,000 km per second and (b) physical space manifests the finest fractures as elementary particles, smallest entities in the universe. We know, the more rigid an object, the finer are the results of fracture. Glass panes fracture more finely than wooden planks.

Let us look for even more rigid contents in consciousness, more rigid than the physical space as stated above. Concepts are some of the contents of consciousness. We may consider rigid concepts to provide structural stability to consciousness in the face of eternal mutability of phenomena in the consciousness.

For example, some of the most rigid concepts are the concept of integer numbers.

Concepts which help to, tend to depart from the rigidity of integer numbers, for example, limits ($x \to 0$ for instance) convergence of series, inexactness of differentials, un-manifest values of states in quantum mechanics where measurements collapse the state functions to eigenvalues of corresponding operators.

Within the consciousness, the conceptual rigidity of the number '1' (one) tends to impede precipitation into the 'many'. The consciousness attempts to overpower this rigidity of the concept of 'one'. In the process it, the consciousness, gets itself fractured, precipitating in to myriad nodes of ego.

Ekaoham, bahushayyam

This is in a verse from the Chhandogya upanishad, although it's not written in exactly the same way:

> Tad aiksata, bahu syam prajayeyeti, tat tejo srjata: tat tejo aiksata, bahu syam prajayeyeti, tad aposrjata, tasmad yatra kva ca socati svedate va purusah tejasa eva tad adhy apo jayante.
> Chhandogya upanishad: Pt 6. Ch 2. Sh 3

(TRANSLATION
May I see the drama of my own manifestation.)

In a featureless self, out of boredom, ennui, a feeling of listlessness and dissatisfaction, arising from a lack of occupation or excitement, a primal desire, a primal want arises for experiences other than the monotony of self-hood, to enjoy and to celebrate (aananda and utsava; the Sanskrit word 'Utsava' comes from the word 'ut' meaning 'rise' and 'sava' which means 'worldly sorrows' or 'grief'.

Entire consciousness is play of energy and information.

The space harbors potential energies in the forms of various fields as want-fields (say, as electric field energy in capacitors, $E = \frac{1}{2} CV^2$, or, rather, electromagnetic field energy density, $u_{(EM)} = \frac{1}{2} \varepsilon E^2 + \frac{1}{2} \mu B^2$) and (plastic) forms of precipitates (say, $E = mc^2$). The precipitates manifest kinetic energies of precipitates of ego (think of $E = \frac{1}{2} mv^2$ for masses) in their movements in response to the want-fields.

The experiences (exchange of energies and information) are often oscillatory, repetitive from exchanges of energies at various levels.

By the contracting principle of ego multiplicity precipitates in the consciousness.

Story of amoeba

God said to amoeba when he created it, to go and multiply. But amoeba said: "no God, I do not know how to multiply, I only know how to divide". God, who always has the last laugh, said "ok, now go and multiply by division".

A MATHEMATICAL BASIS FOR HOMOEOPATHY

(This chapter is included as it is on a popular topic and is mildly mathematical)

Homoeopathic potencies, when plotted against a logarithmic scale, popularly called a log-scale, yield essentially a straight line. Logarithm, exponential, golden ratio, Fibonacci series are all intimately connected with the shapes of flora and fauna.

Further, when a plant or an animal, a multi-tier stable system, is sought to be restored to a healthy state from a diseased state by homoeopathic treatment, this appears to be an instance of modern feedback theory.

And It Appears Thus…

It is presumed in this chapter that homoeopathy works. We attempt to understand the workings of homoeopathy from three stand points: (i) the logarithmic aspect of life and (ii) the stability from the negative feedback theory, (iii) the significance of repetition in life processes.

Logarithmic plot of homoeopathic potencies

The commercially available centesimal potencies are: 3, 6, 100, 30, 200, 1000, 10,000, 50,000, 100,000.

Plotted in steps against a log-scale in blue color, we find essentially a straight line:

A Matematical Basis for Homoeopathy

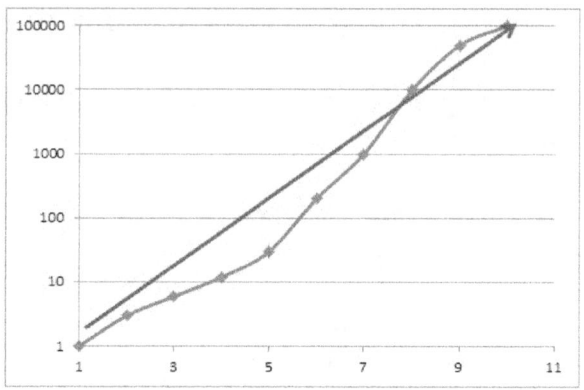

In the figure above, the blue line is a step-wise plot of commercially available homoeopathic potencies. The red line is the idealized straight line plot.

We may infer from the figure above that the structure of vitality in living things has an exponential or, saying the same thing in an inverse manner, a logarithmic structure.

Avogadro's number and the problem of homeopathic dilution

There is great controversy on the claims of homoeopathic potentization.

> *Serial dilution of a solution results, after each dilution step, in fewer molecules of the original substance per liter of solution. Eventually, a solution will be diluted beyond any likelihood of finding a single molecule of the original substance in a liter of the total dilution product.*
>
> [Wikipedia]

A Matematical Basis for Homoeopathy

The molar limit

If one begins with a solution of 1 mol/L of a substance, the dilution required to reduce the number of molecules to less than one per liter is 1 part in 1×1024 (12C) since:

$6.02×1023/1×1024 = 0.6$ molecules per liter

Homeopathic dilutions beyond this limit (equivalent to approximately 12C) are unlikely to contain a single molecule of the original substance and lower dilutions contain no detectable amount.

Yet we read in homeopathic literature about distinct behavior of homoeopathic drugs at different potencies:

For example, in the book 'Select Your Dose and Potency' by Dr. PS Rawat, B Jain Publishers (P) Ltd, New Delhi; page 272, Medicine: Hepar Sulphuris Calcareum, Point 2, we read:

'The higher potencies may abort suppuration, the lower promote it.

Feedback theory

Homeostasis or homoeostasis (homeo + stasis) is the property of a system in which variables are regulated so that internal conditions remain stable and relatively constant. Examples of homeostasis include the regulation of temperature and the balance between acidity and alkalinity (pH). Human homeostasis is the process that maintains the stability of the human body's internal environment in response to changes in external conditions. Homeostasis requires a sensor to detect changes in the condition to be regulated, an effector mechanism that can vary that condition, and a negative feedback connection between the two.

[Wikipedia]

If we, now, consider a healthy human being (or animal or plant) to be a multi-level stable system with body, sensoria, brain, mind, vital, spiritual levels as progressively 'higher' levels. Clearly the concepts of the feedback theory can be qualitatively applied which explains stability. Homoeopathy, thus, may be seen as an extension of the concept of homeostasis to include the higher levels of mind, vital and spiritual levels and the stability of the complte system in the face of disease (dis-ease, disturbance).

Now let us look at logarithm and life.

Logarithm in life and existence

Many life forms have logarithm built in to their physical structure. Nautilus, for example, is one.

Many forms in nature, like, for example, cyclonic storms and galaxies of stars are expressed as logarithmic spirals:

A Matematical Basis for Homoeopathy

Cutaway of a nautilus shell showing the chambers arranged in an approximately logarithmic spiral.
[Wikipedia]

Romanesco broccoli, which grows in a logarithmic spiral
[Wikipedia]

A Matematical Basis for Homoeopathy

An cyclone over Iceland shows an approximately logarithmic spiral pattern
[Wikipedia]

The arms of spiral galaxies often have the shape of a logarithmic spiral, here the Whirlpool Galaxy
[Wikipedia]

Reproduced from an earlier chapter: 'Physics of the Senses':

Logarithmic response of the senses to stimuli

Consider the responses of eyes, ears to brightness or volume of sound – hence the use of the unit of decibel for sound volume. Decibel is a logarithmic unit.

The following is from Wikipedia:

The eye senses brightness approximately logarithmically over a moderate range and stellar magnitude is measured on a logarithmic scale. Perceived loudness/brightness is proportional to log of actual intensity measured with an accurate nonhuman instrument.

[Wikipedia]

Logarithm is intimately connected with life. Homoeopathy is an instance of it.

Repetition and the life process

The practice of repetition is intimately connected with the art of improving life. Asana, pranayama, japa or chanting with or without rosary are made perfect by repetitive practice and improve life. Krishna says *'Yajnaanaam japayajno'smi'* (repetition of mantra is regarded as the best of all sacrifices) in the Gita 10:25.

The Sanskrit definition for mantra is *'mananat-trayate iti mantrah'*— meaning mantra is that which, through constant repetition, or *mananam*, frees one is from the plagues of the world process. Repetition of effort or experience is said to assist in hardwiring in the brain:

> *During the process of solidifying a memory repetition is key, because when an action is repeated the action is engraved into the brain for a short amount of time. The more times it is repeated the easier it will be to solidify the memory.*
> *[Wikipedia]*

> *When an association is made in the brain, groups of adjacent neurons are repeatedly fired up. Through repetition, each time the same groups of cells are activated, their connection becomes more solid until they are physically fired together. This connection is known as 'hardwiring' (view url). It can take up to 2 years to create a connection in long term memory but once it has become solidified in the mind and once this connection is learnt, it is rarely forgotten.*
> *[Wikipedia]*

DEATH WISH OF THE WORLD CIVILIZATION

Some days the movies on almost every second TV channel depicts a death wish for the world: 2012, Armageddon, The day after tomorrow ...the list goes on.

And It Appears Thus...

The world, as a whole, is bracing itself psychologically against a complete destruction of the civilization as we know it. This is a collective psychological reaction of guilt from our collective vandalization of the natural environment.

The roots of this death wish are in the values of ego, competition, exploitation each of us have cultivated as persons, communities and nations. These values might have helped us develop and survive so far, but now we are faced with the threat of collective destruction because of the very same values.

'Exploitation' of limited natural resources, minerals, flora and fauna in equal measure, at an ever increasing rate and profiteering has resulted in inevitable global consequences. Clean and healthy water sources are a thing of the past. So is also clean air.

Those who are in their fifties will remember the clear blue skies of their childhood as a thing of the past. Today, in most cities, the sky has an ash color and the sun looks like a large cinder in a furnace.

More and more action results in ever more number of problems. Sri Aurobindo has said:

> *"All this insistence upon action is absurd if one has not the light by which to act. ----- The advocates of action think that by human intellect and energy making an always new rush, everything can be put right; the present state of the world after a development of the intellect and a stupendous output of energy for which there is no historical parallel is a signal proof of the emptiness of the illusion under which they labour."*

Today you cannot show a living philosopher who can guide the world for a millennium. Nor a poet to give light for a hundred more years in to the future. All the great poets and philosophers we know are from the past.

Free attention hours and culture

A Socrates, in today's materialistic civilization, would be busy producing material wealth serving in an industry or a bank twelve hours a day to sustain him economically or else perish. No more the idyllic life of freedom to gather young men and discuss philosophy while walking the streets. Neither he nor the young people around him would have the leisure to indulge and develop philosophies in today's technological civilization. Free leisure hours are the source of culture. It is culture that supports a civilization. With the culture drying up, civilization will crumble with a time lag. It is instructive to study the GDH (Gross Domestic Happiness) concept of the king of Bhutan in this context.

Why technological advance robs us of free leisure hours?

What is the reason for lessening of personal time as technology advances? This looks paradoxical. However the reason is as follows:

More technology helps build more ways for work to flow, to grow faster than the growth of skilled manpower can cope with; this requires more hours of attention to be paid by every skilled person every day. This is like adding new conductors in parallel to an electric circuit which reduces overall resistance and allows more electric current to flow:

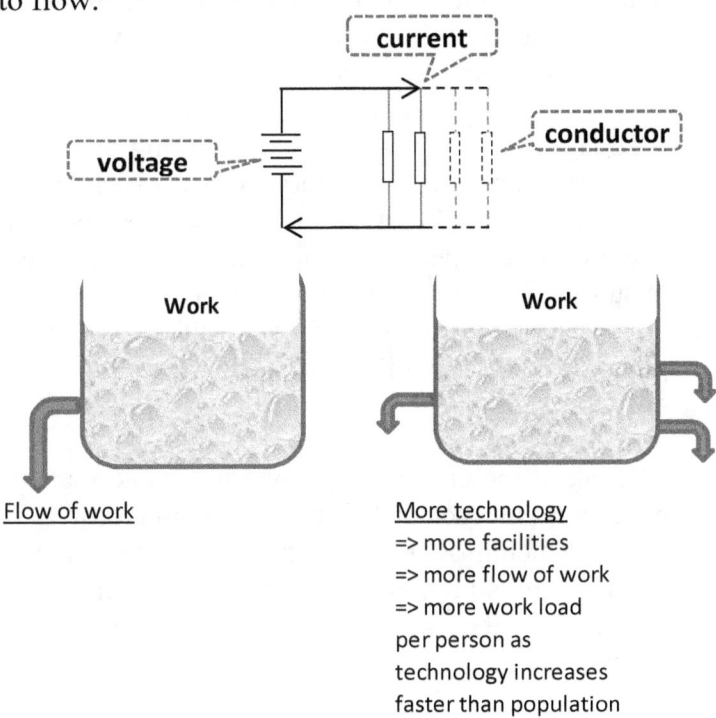

Over-specified Quality of product and the price on environment

When the people-with-money and the experts-with-skills join hands, mute environment pays a heavy price for every dollar spent.

The race for over specifying quality which is unlikely to be of much practical use is an unnecessary crime against mute nature. Natural resources expended on the extra quality may not actually provide a true, higher value of service. Much of the over specified quality of service may be nothing more than fancy stuff.

Today everybody is weary of building and maintaining the vast economic machinery. The very same machinery we build to work round the clock takes its revenge by keeping more people awake at unearthly hours. With leisure gone, even basic science is not developing notwithstanding a 'string' of fruitless speculations in physics: no worthwhile new world-views have manifested after the 1950s.

Think of the enormous growth in literate manpower and especially the scientifically literate manpower since the 1950s. Much growth has been seen in terms of technology, engineering and management; but in proportion to this increase in the scientific manpower, how have world-views grown?

Vertical Pedal Bicycle

How I Got This Idea

Late in my boyhood I came across an article in Science Today, a popular science magazine in India, descrbing a vertical pedal bicycle invented by a student from an IIT. There were two freewheels on either side of the rear wheel. A chain ran on each freewheel. A pedal hung from each chain on a pulley on either side of the bicycle. After being pushed down, the pedals were restored to their top positions using springs.

I had a hunch. A rope hung on a pulley will have one end pulled up when the other is pulled down. Two pieces of chain may be taken, one on each of the two freewheels, fixed on either side of the rear wheel. These two pieces of chain may be connected to three pieces of rope running on three pulleys. The two pieces of chain and the three pieces of rope may be connected alternatingly to make a single length of rope-chain-rope-chain-rope. This arrangement will pull up a pedal when the other pedal is being pushed down. Now no springs are required to restore the pedals to their top positions after being pushed down. Look at the drawing in the following page.

The conventional bicycle with pedals moving in circles is, for most of the circular path, not adapted to utilise the nearly vertical force applied on the pedal. The vertical pedal system seeks to improve the situation by a system of pedals moving approximately vertically and parallel to the applied force.

Dr. G. Rajsekhar, my teacher at IIT-BHU financed this project.

Conventional Pedal System

THE PEDALS in the conventional bicycle move in circles. Only the component of the force perpendicular to the crank shaft helps transfer energy. The component parallel to the crank shaft cannot transfer energy as the crank shaft is a rigid body; there cannot be a movement parallel to the rigid crank shaft (note: energy transferred *or* work done = force * movement along the direction of force).

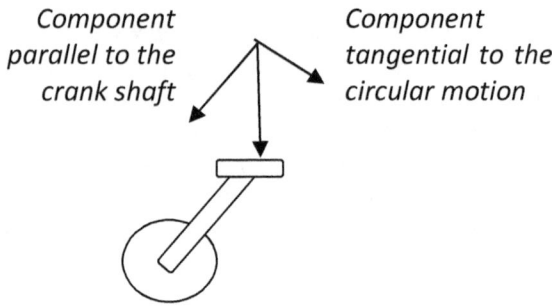

CONVENTIONAL BICYCLE

Vertical Pedal System

The rear wheel is provided with two freewheels, one on either side. An iron rope-chain-string length is arranged suitably to move on the two freewheels and three pullys and is connected to two pedals at either end. When the pedals are pushed alternately down by both feet, the energy is transferred to the rear wheel of the bicycle via the two freewheels stated above. Certificates are displayed at the end of this book.

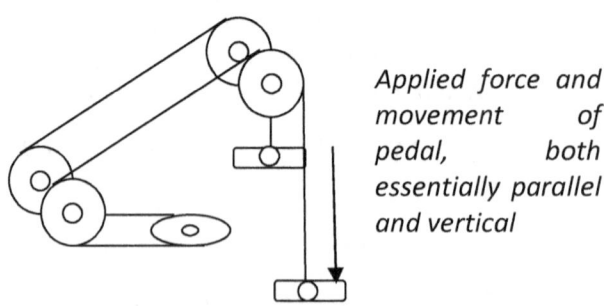

Applied force and movement of pedal, both essentially parallel and vertical

VERTICAL PEDAL BICYCLE

Vertical Pedal Bi-cycle

Finally Speaking

Finally speaking, I feel that the conventional bicycle with pedals moving in circles is a compact arrangement, a very good engineering requirement and is far better than the vertical pedal system. In case of the conventional pedal system, the vigorous pushing is done, instinctively, where the crank shaft is fairly parallel to the ground; both the direction of the push of foot as well as the tangential movement of pedal are downward, parallel to each other and quite efficient.

Transverse Wave Propulsion in Fish and Birds

How I Got This Idea

Near to my home at Chatrapur, Odisha, there is a beautiful lake, Tampara, which I used to frequent and boat across. Once I was intently watching fish trying to figure out how they swam. I ruled out the flimsy fins as propulsion apparatus. Fins may be useful for ensuring stability and control but not for providing propulsion. I observed, every time a fish starts moving, the movement starts with a side wise shake of its head and a subsequent wave in its body which proceeds to the tail. I intuited that the wave profile, while it proceeded from head to tail, pushed the surrounding water backwards. By Newton's third law of motion, the reaction from the surrounding water pushed the fish forward. Later, I took my sister's school box and put some water and a small sheatfish in to it and closely watched the fish swim. After perhaps an hour, I was convinced that I was correct. Subsequently, referring to the Encyclopedia Britannica, I found the very concept was described to explain the motion of fish.

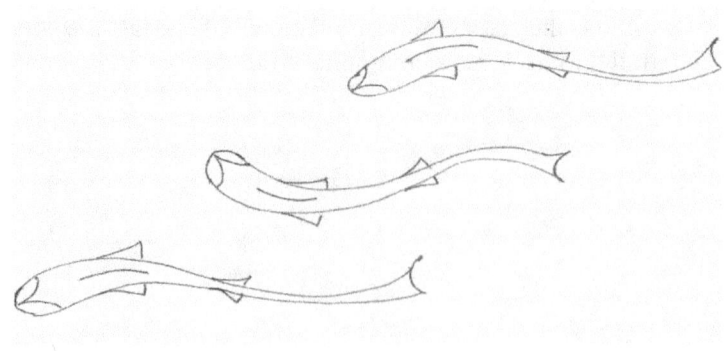

I felt I can simulate the fish mechanically. This I did in the year 1980 in the wave-boat described below. Subsequently I felt, I can create a water pump by encasing the propeller as an impeller. This I demonstrated in 1988 in a wave-pump described later on.

Still later, watching cranes, flying alongside while I was on trains slowly moving, I felt the same principle was at work in the wings of the birds; the only added complication was that the wings must also provide lift from the Bernoulli's principle (in aircrafts, however, the wings provide only lift). I demonstrated the principle of the wave-propeller in air in the year 1993. This is described at the end.

Transverse wave propulsion is fundamental to the propulsion of both fish and birds.

Chronological Development

Wave Boat, Modex- 80, Year 1980

IN THE YEAR 1980 the *Wave Boat* was exhibited in the Model Exhibition *MODEX- 80* in the IIT, Benaras Hindu University. The idea was counter intuitive considering that a cork floating in water will only bob up and down when a wave passes it by. It was a surprise to many including my professors one of whom laughed with glee when the Wave Boat moved in water.

A rubber cloth, held taut by springs, was oscillated up and down by a flywheel-crankshaft mechanism driven by a small motor. Transverse waves were created at the leading edge of the rubber sheet and proceed to the rear edge of the rubber sheet. The water surrounding the rubber sheet was pushed backwards by the profile of the transverse waves. By Newton's 3rd law of motion the boat moved forward. The body of the Wave Boat was in the shape of a pontoon.

The basic idea had been borrowed from fish. After watching carefully fish swimming in a lake I concluded that the fins, being flimsy, were at most useful for stability and control purpose, but not for the purpose of delivering propulsive power. Noticing that there is undulating movement of the body of the fish, starting at its head and proceeding to its tail every time the fish started to swim, suspicion dawned that the wave motion is indeed fundamental mechanism of fish propulsion. That began the Wave Boat experiment. It won a prize among *Advanced Models* category.

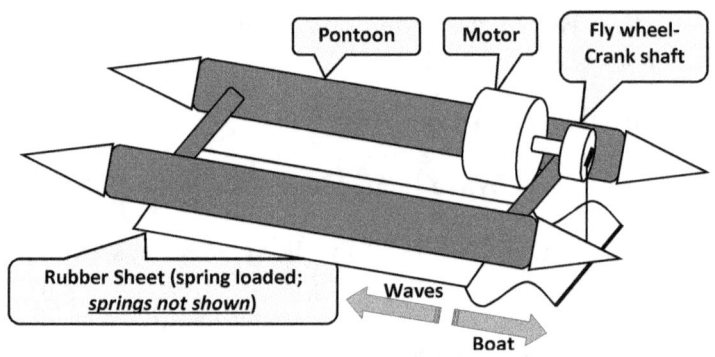

Schematic Diagram, Wave Boat (Propeller)

Wave Boat (propeller), Model Exhibition- 1980, IIT-BHU

Wave Pump, Year 1988

A pump is where the propeller is called an impeller and is confined in a casing that is held stationary. Approximately 6 gallons per minute water issued out of the discharge pipe in a smooth flow. The maximum head at no discharge was 3 meters. The impeller was oscillated about a fixed axle something like a hand fan. This was a practical engineering adaptation of the ideal transverse wave motion.

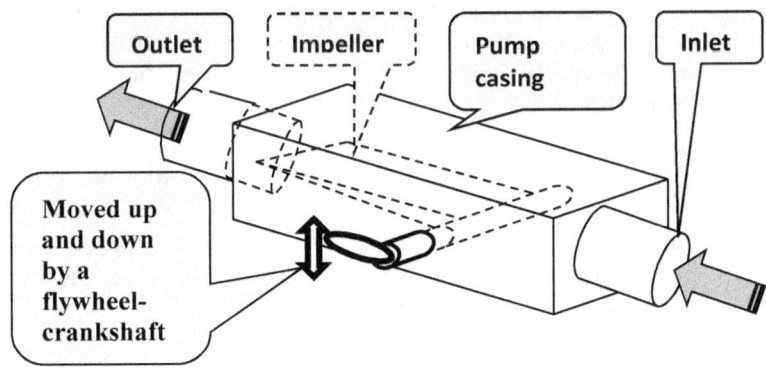

Schematic Diagram, Wave Pump

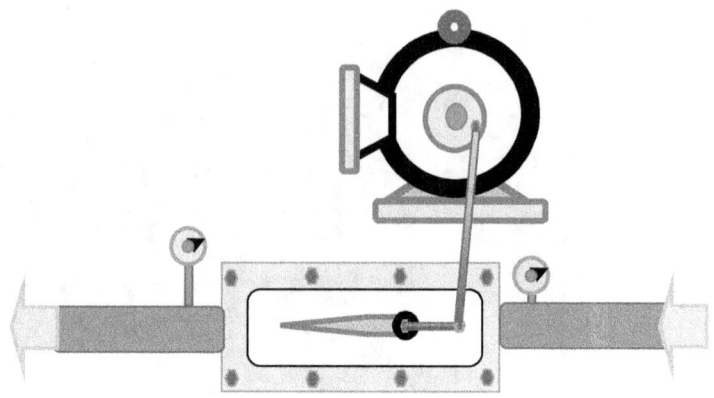

Artist's Impression of the Wave Pump

Wave Propeller in Air, Year 2003

There was no actual flying. *Only the propulsion from transverse waves was demonstrated for the air propeller.* The experiment was similar in concept to the testing of propulsion of a jet engine or a rocket engine on a test bed.

In birds the wings provide both lift as well as propulsion. In fixed wing aircrafts the wings provide lift whereas propeller or jet engine provides propulsion. The purpose of the experiment was only to demonstrate the principle of propulsion in air by transverse waves. When applied to actual aircrafts, fixed wings would provide necessary lift and wave propeller would provide propulsion.

Threads were used to suspend the model aircraft to take care of the need for lift. A small dc motor powered by 1.5 Volts battery was used to drive the propeller. The principle of propulsion in air by transverse waves was demonstrated conclusively. A similar propeller driven by an engine of adequate power will propel an aircraft to flight.

The propeller was oscillated about a fixed axle something like a hand fan. This was a practical adaptation of the ideal transverse wave motion just as in the Wave pump of 1988.

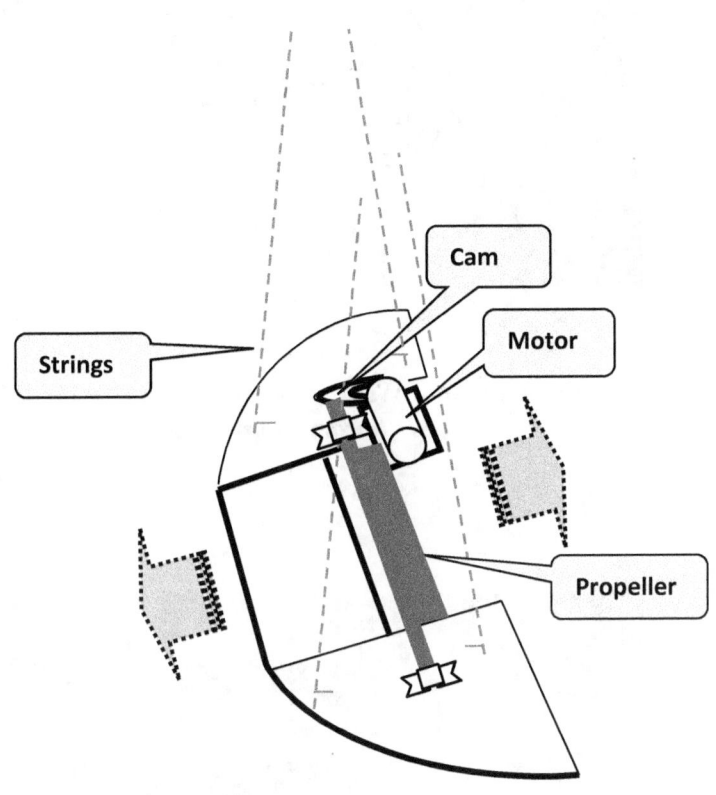

Schematic Diagram of Wave Propeller in Air

Demonstration of Wave Propeller in Air (*in the laboratory-library in my house*)

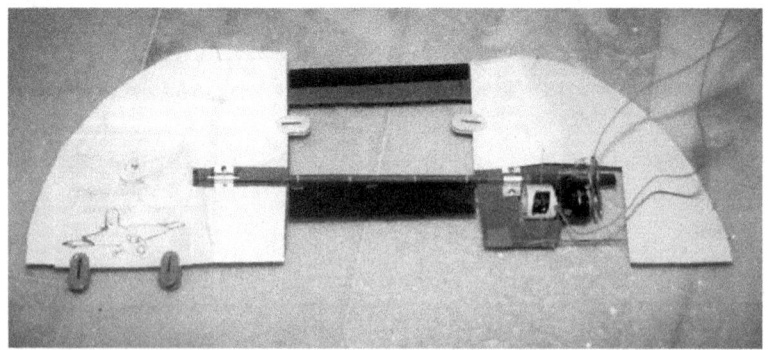

Wave Propeller in Air

Promising Applications

Low Noise Micro Size Aircraft

A low noise, lightweight, low power, slow flying, cheap micro size aircraft may be fashioned around the transverse wave propeller.

A conventional micro size aircraft is propelled by an ordinary propeller that is prone to high noise resulting from tip speed (that is proportional to both rpm and diameter). The transverse wave propeller may overcome the noise problem to an extent by imitating bird flight. The associated loss of energy from noise may also be conserved.

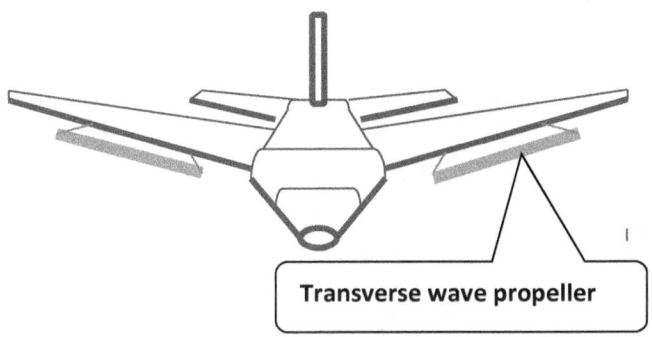

A Failed Experiment with Unpatriotic Fish

I dreamed of domesticating fish which may be encased in low head (say 6" head) 'live fish pumps'. Such 'live fish pumps' may be used in large numbers to pump water in paddy fields from plot to neighboring plot in stages. The fish may get a supply of nutrition from the water flowing through the pump casing. To demonstrate the idea of a live fish pump, I constructed a box with inlet and outlet openings and let water and fish in to the box:

The fish refused to swim. They were unpatriotic. My dream of conserving power and diesel across India's vast countryside was dashed. (Not entirely yet; I still dream of succeeding in demonstrating the idea with the help of ichthyologists and geneticists who can help grow flat muscular bodied fish that quickly grow to adulthood and hold their size steady, pump water indefatigably and, after their useful life is over, provide gastronomic pleasure to the farmer and his family. Any volunteers?)

Building Wound-Core Transformer

How I Got This Idea

While undergoing training at BHEL, Bhopal, as a 4th year engineering student, I was sent to the transformer manufacturing section where very large transformers, some a few hundred MW, were being built. Looking at the core building process which involves a great deal of precision machining, I wondered why is it necessary to expend so much engineering effort on an object which is mechanically static, free of moving parts. There I got this idea. I applied for and got patent rights for the same. I also published an article on the subject in the well known journal, Electrical India. The text of the article is reproduced below:

Beginning of Text:

Building Wound-Core Transformer

by

Ligaraj Patnaik, M.I.E.

Abstract:

A method of building transformers with continuous core is illustrated. This wil have obvious theoretical and practical advantages. Time and cost of manufacturing will be reduced. The manufacturing can be completely machanised.

Concept:

Building the Core

PLEASE REFER Fig. 1. Buildig the core can be accomplished wih a mould having two straight sides. Avoiding sharp bends in the core will help minimise the internal stresses. The mould may be mounted on a horizontal turn table. Care must be exercised to achieve the two straight sides of the core as precisely as possible. Spring loaded idling rollers exerting lateral pressure may be used for the purpose. At the end of the operation tight bracings must be used to maintain the shape.

Conductor Winding

Please refer Fig. 2. For winding the conductors, a splittable mould will be introduced about one straight side of the core. It can be held floating with the help of various rollers. The mould may have, on one or both the circular ends, means to attach a belt, a chain or a gear to drive it. Then suitable conductor can be wound upon it. The mould wil be left behind in the transformer. When the occasion arises for repairing the transformer, the winding can be unwound with the help of this mould. By this means the other winding(s) of either one phase or three phase transformers can be built.

Building Wound-Core Transformer

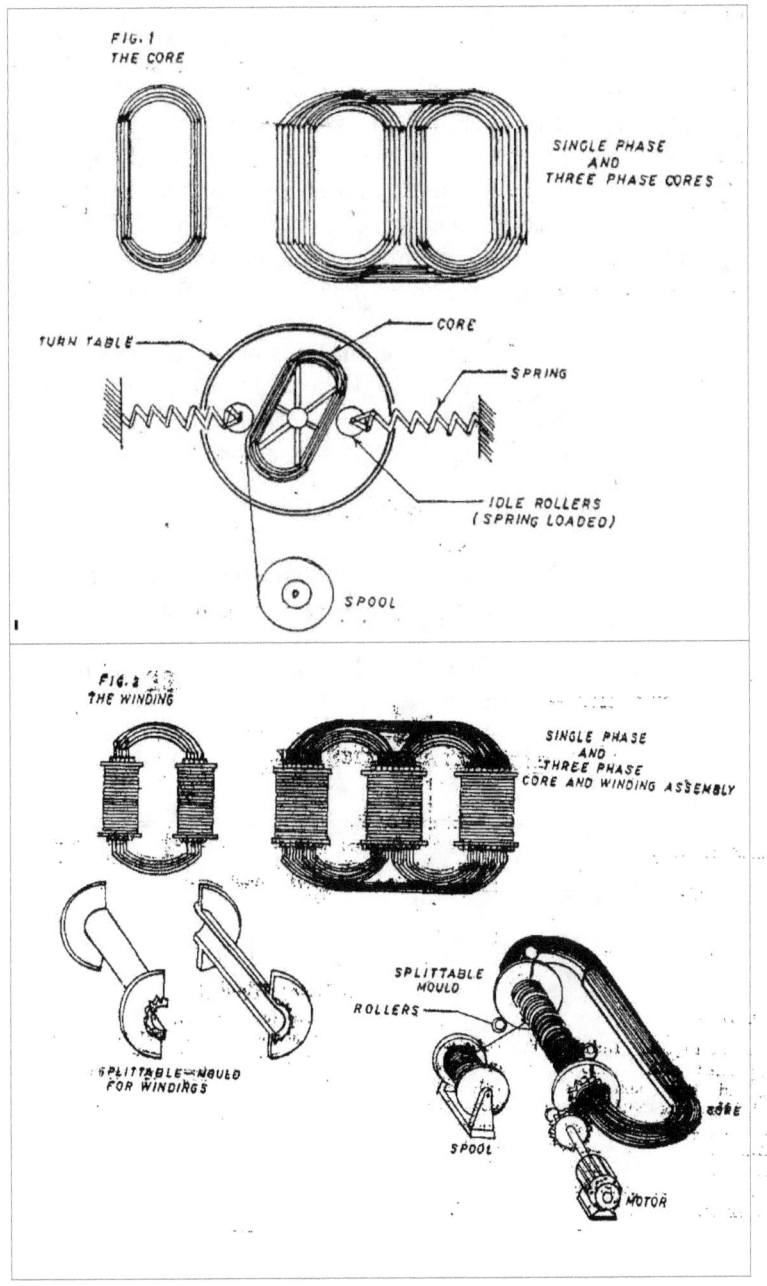

Advantages

Employing this method of construction will help avoid many of the cutting operations like cropping, piercing, mittring. Also this will help avoid much of the manual operations like unlacing, relacing, etc. This method of building can be more easily automated than the conventional one. Besides the above, real benefits like (i) continuous, hence, more efficient core and (ii) less wastage of core material can be obtained.

Leaving behind splittable moulds need not present any special problems. In fact, this will be helpful in unwiding damaged windings and rewinding new one without requiring unlacing and relacing, thus saving on time, effort and cost of repair work.

Resume:

Existing manufacturing lines can be considered for adoption of this method of core building.

Existing transformers, when beyond repair, can be replaced with new ones.

===

End of Text.

Solution to Nuisance Tripping of Plant Lightning from Fault in Appliances

How I Got This Idea

Fed up of me for insisting on resetting the earth leakage circuit breakers (ELCBs) every time these tripped, causing lights and fans to go off, a very good friend of mine threatened to throw me and my table out of the shared work space. That set me thinking for a solution. I felt that ceiling fans and lights need not be covered by 30 mA ELCBs as these are generally handled by professional electricians. Only the appliances plugged to the wall sockets expose the lay users to the hazard of electricity. It is sufficient to cover only these wall sockets by the ELCBs leaving the ceiling fans and lights alone. The reputed journal 'Electrical India' have kindly published an article which is reproduced below.

Beginning of Text:

=======================================

Solution to Nuisance Tripping of Plant Lighting from Fault in Appliances

By
Lingaraj Patnaik MIE

Solution to Nuisance Tripping of Plant Lightning from Fault in Appliances

Nuisance tripping of plant lighting circuits protected by Earth Leakage Circuit Breakers is a nightmare for electrical maintenance people especially when a portion of the plant lighting trips for no better reason than an earth leakage or earth fault in, say, a plug socket outlet.

Whereas safety is achieved upto 100%, circuit availability is reduced below acceptable levels. And this, in many instances, leads to by-passing of the ELCBs by the maintenance people.

The trouble arises because the same ELCB which potects the luminaire circuits also protects the power socket outlets meant for appliances.

Whereas luminaires do not develop earth faults frequently the same is not true for appliances hooked to power socket outlets where frequent earth faults do take place.

The solution to the above problems could be in the following directions:

1) All plug sockets and appliances should be fed from independent MCB DBs to be de-

Solution to Nuisance Tripping of Plant Lightning from Fault in Appliances

signated as 'Appliance MCB DBs.' Only luminaire circuits should be fed from 'Luminaire MCB DBs'. If this is done plant lighting shall not suffer nuisance tripping owing to appliances.

2) If however, the same MCB DB is to be used for both luminaires and appliances then 2 phases should be used for luminaires and the 3rd phase should be reserved for plug sockets and appliances only. Of course plug & sockets should be kept in boards separate from luminaire switch boards.

3) 100 mA/300 mA ELCB may be used for the 'luminaire MCB DBs' or the 'Luminaire phases' (2 out of 3 phases) and 30 mA ELCB for the 'Appliance MCB DBs' or the 'Appliance phase' (3rd phase).

This will ensure fire hazard protection (100/200mA) for the luminaire circuits and personal safety (30 mA) for appliances circuits, and, thus, we feel, will provide the optimum performance for the combined availability and safety aspects.

=====================================

End of Text.

Dogma and the Baconian Filter for Separating the Scientific from Speculative Theories

A Peaceful Evening in My Household

My young daughter and her cousin started a discussion on the solar system one evening at dinner time. We were squatting on the floor, Indian style. On being asked whether the sun revolves around the earth or the earth moves around the sun, both of them, wise in their sixth grade, stated emphatically that the sun stands still and the earth rotates around its axis. Sneeringly I asked, whether all of us appeared to be rotating? Whether the house appeared to be rotating? Or the trees appeared to be rotating? Or were our heads rotating? Visual evidence was against them. Playfully I pointed out that in their third grade they had studied and written in the examinations that the sun rises in the east and sets in the west. So both the teachers and the text books of the third standard were wrong, or else, the teachers and the text books of the sixth standard are wrong. Their pride was pricked. By this time, having finished my dinner, I was sitting on the sofa. Peeved, the children flew at me from either side and tried to convince me, by physically pinning me down, saying that I did not know anything, and threatened to take me to a scientist uncle who shall explain me everything. Subsequently the uncle assured them that the earth was, indeed, rotating around its axis and the sun is still. Point is, without accepting additional physical principles, such as the principle of 'minimum total energy of a system', one cannot determine with certainty which particular version is correct. The simple principle of 'equal but opposite mutual relative velocities' would admit of both the versions as equally correct, leaving the matter to be decided by each observer for himself. However, today, we are conditioned to abandon direct sense evidence lightly in favor of authority.

Dogma and the Baconian Filter for Separating the Scientific from Speculative Theories

CENTURIES OF SCIENCE has not helped us shed our habit of dogma. When one group of gods fall, another takes over and carries on as if infallible. It is time to submit every theory to the Baconian filter, as exemplified below, before we pronounce a theory to be either a scientific or, merely, speculative theory.

Fundamental Laws of Physics

The statement of every fundamental law of physics is a statement of faith. Take, for example, the laws of Newton, Maxwell's equations, postulates of the special and the general relativities Einstein, Schrödinger's equations, Heisenberg's uncertainty principle, etc.

Once a law can be derived from other fundamental laws, it is no more fundamental law, now it is a derived law. For example, the ideal gas laws are derived laws (derived from Newton's laws of motion in statistical mechanics). In order to be considered complete, a treatment of the fundamental basis of any theory must always be accompanied with *a checklist of limitations*.

Example 1

List of Limitations (Baconian Filter) of the Newtonian Mechanics:

	Points of examination	Observations
1	Errors/ difficulties/ paradoxes carried forward from tributary theories	
2	Errors/ difficulties/ paradoxes of precedent theories solved	
3	New errors/ difficulties/ paradoxes created	1. The terms space, time, energy, action, reaction are not defined. 2. The first law is actually a definition of

Dogma and the Baconian Filter for Separating the Scientific from Speculative Theories

			inertia, a property of matter, like any other property of matter, for example, volume, density, etc. Newton, as a matter of fact, speaks of inertia as a definition before he stated it as a law of nature in his *Principia*. 3. The first law, indeed, can be derived as a special case of the second law of Newton for an force $F = 0$.
4	Postulates/ additional list of postulates: tacit ones, generalizations made on the original set of the postulates, that lead to various results not reachable by the first set of postulates		
5	Mathematical liberties taken/ limitations accepted		
6	Claims		
7	Disputes		
8	Experiments proposed for verification of the predicted results and the extent to which the theory shall stand verified		
9	Extent of experimental verification, repeatability and ranges of various errors		

10	Possible alternative theories/ suggestions/ recommendations	
11	Examination of theoretical contribution in terms of	
	(a) Simplification of method	
	(b) Simplification of presentation	
	(c) Revision of existing fundamentals	
	(d) Further development of existing theories	
	(e) Consolidation of different existing theories	
	(f) Mere new definitions, symbols	
	(g) Mere indulgence in speculations which cannot be or has not been verified	
12	Practical applications	

I have left gaps which the reader may attempt to fill.

Example 2

List of Limitations (Baconian Filter) of the Theory of Electromagnetism of Maxwell:

	Points of examination	Observations
1	Errors/ difficulties/ paradoxes carried forward from tributary theories	
2	Errors/ difficulties/ paradoxes of precedent theories solved	1. Electricity and magnetism were

		combined. 2. c = constant was derived for waves.
3	New errors/ difficulties/ paradoxes created	4. Redundancy: There are two unknowns **E** and **B** but four equations: The two *divergence* equations can easily be derived from the two *curl* equations. 5. The physical reality of the Poyinting vector is not established.
4	Postulates/ additional list of postulates: tacit ones, generalizations made on the original set of the postulates, that lead to various results not reachable by the first set of postulates	
5	Mathematical liberties taken/ limitations accepted	1. Lack of mathematical methodology forcing the choice of separation of variables by way of the (Ludwig) Lorenz gauge.
6	Claims	
7	Disputes	1. Lack of ready compatibility between Maxwell's equations and $E = h\nu$ for photons.
8	Experiments proposed for verification of the predicted results and the extent to which	

		the theory shall stand verified	
9		Extent of experimental verification, repeatability and ranges of various errors	
10		Possible alternative theories/ suggestions/ recommendations	
11		Examination of theoretical contribution in terms of	
		(a)　　Simplification of method	
		(b)　　Simplification of presentation	
		(c)　　Revision of existing fundamentals	
		(d)　　Further development of existing theories	
		(e)　　Consolidation of different existing theories	
		(f)　　Mere new definitions, symbols	
		(g)　　Mere indulgence in speculations which cannot be or has not been verified	
12		Practical applications	Endless.

Example 3

List of Limitations (Baconian Filter) of the Theory of Special Relativity:

	Points of examination	Observations
1	Errors/ difficulties/ paradoxes carried forward from tributary	

		theories
2	Errors/ difficulties/ paradoxes of precedent theories solved	1. Explained constancy of the velocity of light from the Michelson- Morley experiment.
3	New errors/ difficulties/ paradoxes created	1. Either the special theory of relativity, which asserts that no finite rest mass can move at the speed of light, is correct or else, the traditional rule of 'equal but opposite mutual relative velocities between any two objects' is correct because I routinely move at the velocity of light past every photon streaming in my neighborhood.
4	Postulates/ additional list of postulates: tacit ones, generalizations made on the original set of the postulates, that lead to various results not reachable by the first set of postulates	1. c = constant. 2. The generalization of $dr^2 - c^2 dt^2 = 0$ to $dr^2 - c^2 dt^2 = ds^2$ where ds is defined as a space-time interval and postulated as an invariant is a stroke of intuition of Einstein. This may be seen as the central postulate of the special theory of relativity from which the facts (i) c = constant and (ii) the Special (Hendrik) Lorentz Transformations can be derived.

Dogma and the Baconian Filter for Separating the Scientific from Speculative Theories

		Bypassing this postulate, the Special (Hendrik) Lorentz Transformations cannot be derived from c = constant. Thus, $dr^2 - c^2 dt^2 = ds^2$, with the advantage of hindsight, may be seen as the central postulate of the special theory of relativity.
5	Mathematical liberties taken/ limitations accepted	
6	Claims	
7	Disputes	
8	Experiments proposed for verification of the predicted results and the extent to which the theory shall stand verified	
9	Extent of experimental verification, repeatability and ranges of various errors	
10	Possible alternative theories/ suggestions/recommendations	
11	Examination of theoretical contribution in terms of	
	(h) Simplification of method	
	(i) Simplification of presentation	
	(j) Revision of existing fundamentals	
	(k) Further development of existing theories	

	(l) Consolidation of different existing theories	
	(m) Mere new definitions, symbols	
	(n) Mere indulgence in speculations which cannot be or has not been verified	
12	Practical applications	

A Dream By Way Of Conclusion

I dream of a large world library for fundamental theories, built with a tunnel like architecture, where theories dealing only with the foundations of physics find place. Only those theories that pass the Baconian filter will find entrance in to the sanctum sanctorum of 'scientific' theories.

No doubt, refinements shall be made to the Baconian filter from time to time. No theory will stand the scrutiny of one or all the versions of Baconian filters always. Consequently all the theories are apt to be classified as part scientific and part speculative at any given moment of time. Besides, the candidate theories may be shuffled back and forth between the two chambers from time to time and this shall represent the vigor of the current scientific activity. We must never forget that where as a collection of facts of nature comprises a discovery, a theory is always an invention of the human mind. No theory, hence, shall attain the status of a perfect scientific theory forever.

Dogma and the Baconian Filter for Separating the Scientific from Speculative Theories

The Dream Library:

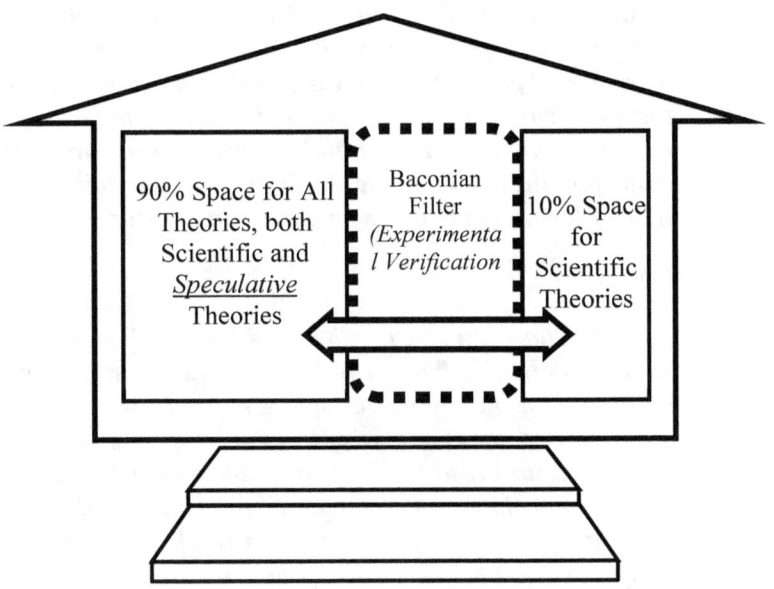

A Conceptual Error in the Calculation of Length Contraction in Special Relativity

(An exception is made in including this mathematical chapter in this book of non-mathematical chapters because this chapter elucidates the basic ideas on special relativity which are popular. Special relativity is very simple mathematically- high school geometry and algebra shall mostly suffice- though the physical ideas are extremely difficult to grasp. General relativity, on the other hand, is very simple in its physical concepts but is extremely difficult in its mathematical make-up employing curvilinear geometry and tensor calculus.)

Core Idea

By the time you are reading this line, its image is already 2 nanoseconds old. When I notice a friend eight feet away he is already eight nanoseconds older. When I hold a foot scale in my hand and see it edgewise I not only see a span of space one foot long, but I also 'see' an associated span of time equal to a nanosecond. Indeed I suffer an illusion when I think that I see the entire foot scale as it exists at a single moment of time. The image of the far end of the foot scale reaching my retina is necessarily a nanosecond older than the image of the near end of the foot scale reaching my retina. This is a fact of physics. To stretch the point a little farther, it is indeed true to say that, in the event the far end of the foot scale explodes, I will notice the event only after a nanosecond later. However, for the duration of a nanosecond, I will be thinking that the foot scale is intact. This is because the image of the far end seen at the moment is a historical image. (If it is difficult to associate nanosecond responses with physiologically slow eyes, think in terms of extremely fast light sensing devices). My perception of a span of space is always, necessarily, accompanied with a historical span of time. This is the first of the two influences of time on our perception of space. The second influence of time on our perception of space arises from bodies moving at great relative speeds comparable to the speed of light.

A Conceptual Error in the Calculation of Length Contraction In Special Relativity

We shall explore the core idea of this chapter after a small detour.

For a very long time I have wondered about my relative speed with respect to the photons streaming past me at 300,000 km per second in my neighborhood. I cannot see how it can be anything different from the same 300,000 km per second but in the opposite direction. After all, if P goes at a speed v with respect to Q, then Q must go at the same speed v with respect to P but in the opposite direction. Clearly either this principle of 'mutually same relative speed between any two observers but in opposite directions' is wrong or else special relativity is wrong. Both are not compatible with each other. After all the photons are physical objects, packets of energy, moving about in the physical space in my neighborhood. After this detour, let us go back to the core idea of this chapter.

First we shall set forth the core idea of this chapter in full. Later we shall elaborate the details at leisure. Refer the figure with two complex frames placed at an angle ψ between them.

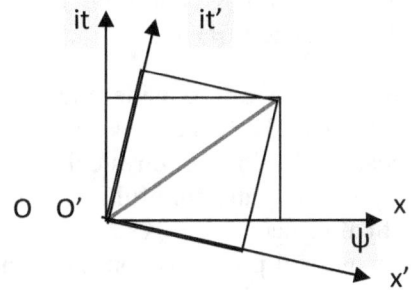

Complex space-time (x, it) and (x', it')

The primed frame (x', it') is the proper frame of a rigid rod and a clock. The unprimed frame (x, it) is the proper frame of

A Conceptual Error in the Calculation of Length Contraction In Special Relativity

an observer. The velocity of light is $c = 1$ unit. We write $\gamma \equiv \sqrt{[1 - v^2]}$ where v is expressed in units of c.

The familiar (Hendrik) Lorentz transformation for length is: $\Delta x = \gamma(\Delta x' + v\Delta t')$ i.e., $\Delta L = \gamma(\Delta L' + v\Delta T')$.

The line of argument in calculating the length contraction is this: at definite time $t = 0$ in the unprimed frame, the position of the integer $x' = 1$, with respect to the unprimed frame, is given by $x = \sqrt{(1 - v^2)} < 1$ implying contraction of length. This is popularly expressed as $L' = \gamma L$ meaning $L < L'$ as γ is always greater than one.

This line of argument is flawed. Let us analyze the matter in light of (a) the familiar transformation of length mentioned a while ago: $\Delta L = \gamma(\Delta L' + v\Delta T')$ and (b) the physical fact that my perception of a span of space is always, necessarily, accompanied with a historical span of time.

The integer $x' = 1$ is the point $(x', it') = (1, i0)$ in the primed frame.

The position $x = \sqrt{(1 - v^2)}$ at time $t = 0$ is the point $(x, it) = (1, i0)$ in the unprimed frame.

Both these points are outside the light triangles in their respective frames. By themselves, these two points represent nothing meaningful in the physical space-time. In order to represent a physically meaningful interval in the physical space, a mathematical projection on x-axis must always be associated with a mathematical projection on the it-axis such that $|\Delta t| \geq |\Delta x|$. Remember, $c = 1$.

Indeed, both these points in the respective complex spaces are in the region, outside the light triangles, which is named as 'elsewhere'. 'Nowhere-never' would also be a correct description. Every point inside the light triangles shall satisfy the condition $|\Delta t'| \geq |\Delta x'|$ or $|\Delta t| \geq |\Delta x|$ in order to meaningfully

A Conceptual Error in the Calculation of Length Contraction In Special Relativity

correspond to, say, a rigid rod, its span of length complete with its associated span of time, in the physical space-time.

Yes, x' = 1 is a mathematical projection on the x'-axis but, it is NOT a physical entity. A mathematical projection can be outside the light triangle, but a physical entity is always within the light triangle, inalienable from its associated mathematical projection along the imaginary time axis. For the rigid rod the projection on the time axis must be $|\Delta t'| \geq |\Delta x'|$ i.e., $|\Delta t'| \geq 1$. Thus, for the observer in the unprimed frame, the mathematical projection, $x = \sqrt{(1 - v^2)}$, represents contraction of only the mathematical projection x' = 1 and NOT that of a physical entity, like a rigid rod, complete with its projection of the associated span of time, t' = 1.

The correct equation for Δx must, at once, involve both the $\Delta x'$ and $\Delta t'$ components, i.e.,

$$\Delta x = [\Delta x' + v\Delta t']/\sqrt{[1 - v^2]} \text{ i.e.,}$$

$L = \gamma (L' + vT')$. We must recognize that T' cannot be dissociated from the rigid rod in the primed frame.

This is the core idea of this chapter. The rest is an elaboration and finding some consequences.

Physical Space

A mathematical space is useful, for example, to make an idealized but an approximate model of the physical space. The physical space exhibits certain unique features unlike a mathematical space. Fixed laws of nature operate in the physical space. Time passes in itGravitational and other fields manifest in it. These various fields possess energy in the physical space. The physical space harbors matter. Electromagnetic waves pass through it at the fixed speed of 300,000 km per second. All these features are natural to the physical space. But these features are optional to a mathematical space.

A Conceptual Error in the Calculation of Length Contraction In Special Relativity

Mathematical Spaces

One is free to propose arbitrary rules of game in a mathematical space (whereas the rules of the game in the physical space are fixed and are the objects of scientific discovery). Passage of time is not represented in a mathematical space until we attach a time-axis to it, at least implicitly. Fields, energy, waves or matter are added symbolically as the need may arise. One may freely speculate that for a given mathematical space $F = ma^2$ (whereas the rule $F = ma$ is the fixed rule that governs dynamics in the physical space).

Classical Space: Mathematical Model of the Physical Space as Conceived in Classical Physics

Space in classical physics is conceived as a three dimensional entity. A universal time flows in it uniformly everywhere. This flow of universal time is conceived to be independent of space. We shall refer to such a space as the 'classical space' represented with the axes x,y,z perpendicular to each other:

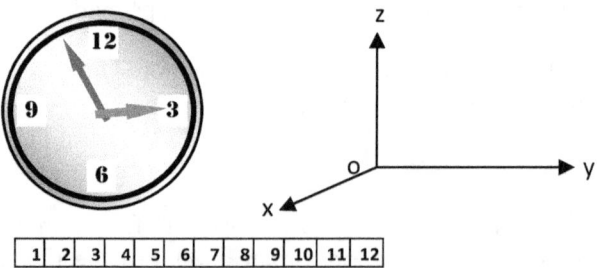

A Complex Space: A Mathematical Model of the Physical Space-Time as Conceived in Special Relativity

Two complex frames are used together. One: the primed frame, (x', ict') for the proper frame of a rigid rod and a clock. Two: the unprimed frame, (x, ict) as a proper frame for an observer. Here c is the speed of light, 300,000 km per second and $i \equiv \sqrt{(-1)}$. These complex frames are placed at a strange angle ψ clockwise to represent a relative velocity v of the rigid rod and clock with respect to the observer (tan ψ = v/ic).

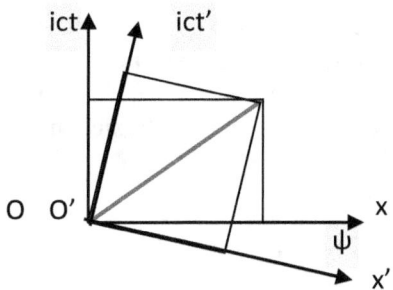

Complex space-time (x, it) and (x', it')

First Limitation Imposed by Time on Our Perception of Space

Light travels at a fixed speed of 300,000 km per second which is same as 1 foot in a nanosecond, or one meter in 3 nanoseconds, or 300 meters in a microsecond or a kilometer in 3.33 microseconds.

A moment's reflection tells us that our perception of a length in reality is necessarily accompanied with a component of time. The star Betelgeuse we see in the Orion constellation in the night sky is already 640 years older. The sun we see in day time is already 8 minutes older. The moon we see is already 1.5 seconds older. When I notice you eight feet away, you are already 8 nanoseconds older. When I see along a bridge one kilometer long, the perception of the length is always accompanied with the passage of 3.33 microseconds. When I see the fingertips at the end of my stretched hand, the finger tips are already three nanoseconds older. When I hold a foot scale in my hand and see it edgewise I not only see a span of space twelve inches long, I also 'see' a span of time a nanosecond long. That I do not realize this fact is an illusion. But the image

A Conceptual Error in the Calculation of Length Contraction In Special Relativity

of the far end of the foot scale reaching my retina is a nanosecond older than the image of the near end of the foot scale reaching my retina. This is a physical fact. My perception of a span of space is always, necessarily, accompanied with the perception of a historical span of time. This is the *first* of the two influences of time on our perception of space. Perception of space is always inseparable from perception of an accompanying component of historical time whether we ordinarily realize this fact of physics or not.

The Physical Space-Time:

Every perception of the physical space by an observer is historical; this is the first of the two limitations imposed by time on our perception of space. Note the (-) signs in the figure below:

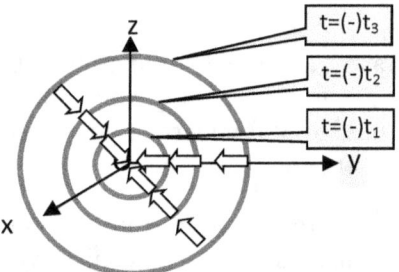

Classical space (x,y,z)

The light spheres carrying historical signals are collapsing on the observer at a speed of 300,000 km per second.

We can represent the above details in a simpler looking mathematical model but which is a little more sophisticated. We generally pay the price of sophistication for making mathematics a little simpler. We express the velocity of light, 300,000 km per second, as a unit of speed, $c = 1$. Now t shall represent a length ct, and v will represent the ratio of the velocities v/c, a unit less number (a ratio of two velocities).

A Conceptual Error in the Calculation of Length Contraction In Special Relativity

The (x, it) **Mathematical Space-Time:**

We construct a mathematical space to represent the physical space with a fixed speed of light. The mathematical space is constructed with two Cartesian coordinates: (i) an axis-x for space and (ii) an axis-it for time. The space and time axes are orthogonal to each other. The space axes y and z are treated as suppressed. An observer is placed at the origin O of the (x, it) coordinates:

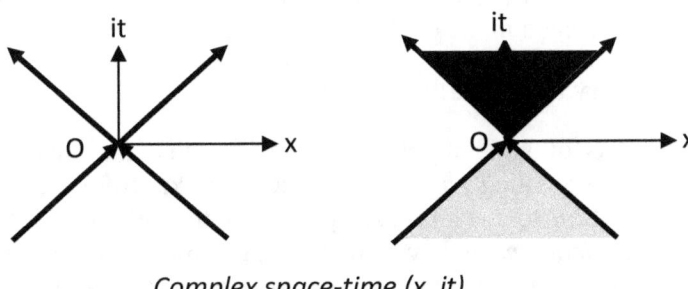

Complex space-time (x, it)

We do not perceive the future. Our world experience is entirely confined to the past. It is, hence, a little more meaningful to draw the complex plane as a single light triangle in the past:

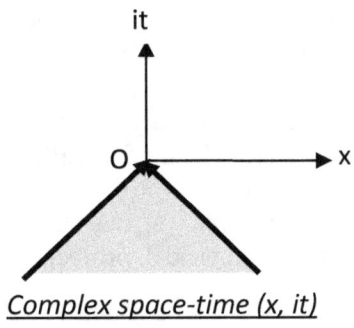

Complex space-time (x, it)

The shaded portion in the bottom half of the complex space-time (x, it) corresponds to the whole of the physical world of experience in the classical space (x, y, z). This, at any moment of time, is composed entirely of the experiences of the past.

This mathematical space (x, it) has some interesting peculiarities. The entire world of experience of the observer in the physical space is contained within the vertical spaces between the two 45^0 lines for which $|\Delta t| \geq |\Delta x|$. This mathematical space, corresponding to the world experience, is the vertical colored portions which we shall refer to as the light triangles and the borders of the light triangles as the light lines. The light lines are the trajectories of light in the space-time represented in the mathematical space above.

A very important realization:

It is of vital importance to realize that the phenomenon of light travelling along the physical x-axis from the positive x-direction to the observer placed at the origin (see the figure *Classical Space (x,y,z)* in the page -62-) is equivalent to the following three components of the complex (x, it) space (see the figure *Complex space-time (x, it)* above) taken together:

(i) the mathematical x-axis in the complex space,
(ii) the 45^0 line in the right side from the past and
(iii) light travelling along this 45^0 line.

Peculiarity-One of the (x, it) Mathematical Space:

No single point of the physical space is represented outside the light triangles of the mathematical space above. Indeed the familiar x-axis of the classical space, a portion of the physical space, perceived by an observer at the origin cannot lie outside the light triangles, certainly not along the x-axis of the (x, it) mathematical space. Yes, what falls on the x-axis of the (x, it) space is a mathematical projection of the perceived x-axis of the classical (x,y,z) space. But this projection is just a mathematical, geometrical projection; it does not represent a

A Conceptual Error in the Calculation of Length Contraction In Special Relativity

physical entity, a physical length along the x-axis in the physical space . A mathematical projection can be outside the light triangle. But a physical entity is always within the light triangle. We must realize that the possibility of the observer at the point of origin somehow reaching out in to this outside-light-triangle-region by using a rigid rod is a big no-no, the possibility does not exist. Super luminal speeds are required for moving a rigid rod from inside to the outside of the light triangle. Infinite forces and energies shall be insufficient to achieve this purpose.

Peculiarity-Two of the (x, it) Mathematical Space

Another peculiarity of this mathematical space is an asymmetry between x-axis and it-axis : that the it-axis lies entirely within the light triangles whereas the x-axis is entirely outside except for the point of origin. An implication is this. The observer at the origin of (x, it) coordinates can perceive the passage of a *'pure'* duration of time, unaccompanied with a space projection. However, he cannot experience a *'pure'* span of length unaccompanied with duration of time. This is because, to be part of the observer's world experience, the span of length must lie within the light triangles which is possible only in accompaniment with a duration of time such that $|\Delta t| \geq |\Delta x|$; this condition must hold for a legitimate space-time interval.

Peculiarity-Three of the (x, it) Mathematical Space

Time changes in an unidirectional manner but space can be traversed in either direction for each of x, y, z axes. The space and the time projections, however, adjust and exchange with each other in order to keep the space-time interval an invariant amongst inertial observers.

Indeed, in writing $(-)\Delta x$, the negative sign is merely an algebraic convention. This does not affect the physical significance of space. One may travel from east to west as easily from west to east. However, in case of time, $(-)i\Delta t$ is not merely an algebraic convention but necessarily represents the physical fact that the duration of time experienced always belongs to the past and never to the future. One can go only in to the future and never in to the past.

Peculiarity-Four of the (x, it) Mathematical Space

The apparent length $\Delta x^2 + \Delta t^2$ of an interval is quite different from its actual length calculated in the complex plane: $\Delta x^2 + \Delta (it)^2 = \Delta x^2 - \Delta t^2$. Indeed, for a space-time interval lying on a light line, i.e., along a 45^0 line, its length in the complex plane is zero irrespective of its apparent length because $\Delta x^2 + \Delta (it)^2 \equiv \Delta x^2 - \Delta t^2 = 0$. One must always check by calculation the length of a line in the complex space and not get carried away by the apparent length.

In line with the last section we may write $\Delta x^2 + \Delta (-it)^2 = \Delta x^2 - \Delta t^2$.

The Central Postulate of Special Relativity

Starting with the postulate of the constancy of the speed of light in special relativity and making use of the generalization of $dx^2+dy^2+dz^2-c^2dt^2 = 0$ to $dx^2+dy^2+dz^2-c^2dt^2 = ds^2$ leads one to the transformation of coordinates of Hendrik Lorentz. Using c = 1 unit of velocity, the generalized equation is: $dx^2+dy^2+dz^2-dt^2 = ds^2$ where ds is defined as the space-time interval. Note, here t is a distance ct with c = 1.

The central postulate of special relativity is this: the generalized space-time interval between two events (x_1,y_1,z_1,t_1) and (x_2,y_2,z_2,t_2) is the same for all inertial observers: s_2-s_1 or Δs or ds is the same for all inertial observers. The standard language expressing this idea is: the space-time interval, ds, is an invariant for all inertial observers. The values of dx, dy, dz and dt adjust amongst themselves suitably in order to keep the value of the space-time interval ds the same for all the inertial observers.

A Conceptual Error in the Calculation of Length Contraction In Special Relativity

The equations of transformation of coordinates between two inertial observers moving at a very high relative velocity with each other were first given by Hendrik Lorentz.

Hendrik Lorentz Transformation of Coordinates

For making a proper choice of mathematical framework, the two complex frames we express the space-time interval as $dx^2+dy^2+dz^2+(icdt)^2 = ds^2$. Here $i \equiv \sqrt{-1}$.

Now we make (x, ict), i.e., (x, it) our choice of coordinates and place an observer at the origin. For convenience we define the letter $l \equiv ct = t$ and, for further convenience, express all velocities of objects, coordinate frames, etc. in terms of units of c. Thus we express the space-time interval as $dx^2+dy^2+dz^2+(icdt)^2 = dx^2+dy^2+dz^2+(idt)^2 = ds^2$. We define below a complex frame as in the figure:

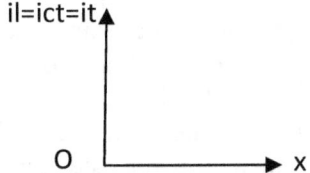

Complex space-time (x, it)

For consistency in our discussions we shall limit ourselves to the pair of primed (x', it') and unprimed (x, it) complex frames.

The primed complex frame is chosen to be the proper frame of a rod and a clock moving at a high relative speed with respect to an observer in the unprimed frame.

The primed complex frame is placed at an angle ψ clockwise with respect to the unprimed complex frame to help us derive the equations of transformation of coordinates. The angle ψ represents the relative velocity in classical space. The idea shall be clear from the relationship $\tan \psi = v/ic$ appearing in the equation (2) coming later on. The origins of the two complex frames are made to coincide.

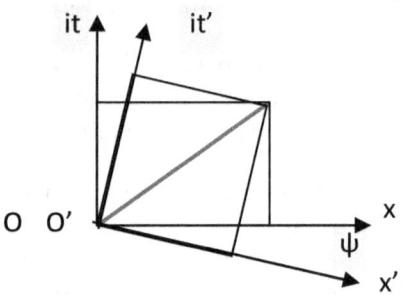

Complex space-time (x, it) and (x', it')

We make two further observations:

1. The rotation of the primed complex plane with respect to the unprimed complex frame is rather a very peculiar idea, even going by the standards of the familiar complex analysis. [Remember the fact that when a complex function $w = u(x,y) + iv(x,y)$ is taken up for the purpose of determining its analyticity, the two complex planes (x, iy) and (u, iv) are always kept apart and never merged together or rotated with respect to each other].

2. In view of the foregoing discussions we must speak of a 'relativistic change of length' always along with the accompanying 'relativistic change of time'. Both the space and time projections must always be required to satisfy the following two cardinal principles:

(i) $\Delta x'^2 + \Delta(it')^2 = \Delta x^2 + \Delta(it)^2$, invariance of interval true for every interval in the complex planes, and,

(ii) $|\Delta t'| \geq |\Delta x'|$; $|\Delta t| \geq |\Delta x|$, true only for legitimate and physically sensible intervals within the light triangles, and not true for the superluminal regions outside the light triangles.

Errors in Calculations of Observed Length in Some Books

A list of errors in various historical, popular and text books is given in the concluding paragraph of this chapter.

Correct Equations

We reproduce the earlier figure below:

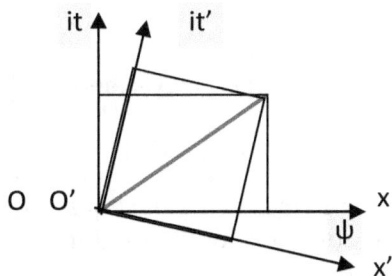

Complex space-time (x, it) and (x', it')

We write the following equations for the two mathematical projections of the rigid rod in its proper frame, the primed frame on x'-axis and it'-axis respectively:

$$\left. \begin{array}{l} \Delta x' = \Delta x \cos \psi - i\Delta t \sin \psi \\ i\Delta t' = \Delta x \sin \psi + i\Delta t \cos \psi \end{array} \right\} \quad (1)$$

We consider a point fixed in the classical (x',y',z') space. For this point, $\Delta x' = 0$ in the (x', it') mathematical space. Thus, from equations (1), we receive:

A Conceptual Error in the Calculation of Length Contraction In Special Relativity

$$\left. \begin{array}{l} \Delta x \cos \psi - i\Delta t \sin \psi = 0 \quad \text{i.e.,} \\ \tan \psi = v/ic \\ \cos \psi = \gamma > 1 \\ \sin \psi = (v/ic)\,\gamma \quad \text{where} \\ \gamma = 1/\sqrt{[1 - v^2/c^2]} \end{array} \right\} \quad (2)$$

Inserting various values from (2) into (1) and, writing velocity v in units of c, and simplifying, we receive the values of the two mathematical projections of the rigid rod as :

$$\left. \begin{array}{l} \Delta x' = [\Delta x - v\Delta t]/\sqrt{[1 - v^2]} \quad \text{and} \\ \Delta t' = [\Delta t - v\Delta x]/\sqrt{[1 - v^2]} \end{array} \right\} \quad (3)$$

We may solve the two equations (3) for Δx and Δt, mathematical projections of the rigid rod in the unprimed frame (x, it) and receive

$$\left. \begin{array}{l} \Delta x = [\Delta x' + v\Delta t']/\sqrt{[1 - v^2]} \quad \text{and} \\ \Delta t = [v\Delta x' + \Delta t']/\sqrt{[1 - v^2]} \end{array} \right\} \quad (4)$$

These are the familiar (Hendrik) Lorentz transformations.

The change of the observed length of the rigid rod from its proper length is the *second* of the double influence of time on perception of space.

Untenable Consequence of Proposing an Unphysical Rod Along the x'-axis (with Zero Component Along the t'-axis):

We may, now, focus our attention on the consequence of the commonplace erroneous proposal of placing a rigid rod on the x'-axis of the primed frame (x', it') and assuming that $\Delta t' = 0$, i.e., erroneously treating the portion from (0, i0) to (x', i0) as

A Conceptual Error in the Calculation of Length Contraction In Special Relativity

a physical interval occupied by a rigid rod. To show the error, we insert $\Delta t' = 0$ in equations (4). We, thus, receive $\Delta x = [\Delta x']/\sqrt{[1 - v^2]}$ and $\Delta t = [v\Delta x']/\sqrt{[1 - v^2]}$. Considering Δt and Δx as positive quantities, and dividing Δt by Δx we further receive, $\Delta t/\Delta x = v < 1$ in units of c, i.e, $|\Delta t| < |\Delta x|$. This space-time interval in the unprimed frame is unphysical.

Summary

We reproduce the equations (4) below:

$$\left. \begin{array}{l} \Delta x = [\Delta x' + v\Delta t']/\sqrt{[1 - v^2]} \text{ and} \\ \Delta t = [v\Delta x' + \Delta t']/\sqrt{[1 - v^2]} \end{array} \right\} \quad (4)$$

We write L for the mathematical projection Δx, T for Δt and similarly L' and T' for the primed mathematical projections $\Delta x'$ and $\Delta t'$ and receive:

$$\left. \begin{array}{l} L = \gamma(L' + vT') \quad \text{and} \\ T = \gamma(vL' + T') \end{array} \right\} \quad (5)$$

Inverting, we receive

$$\left. \begin{array}{l} L' = \gamma(L - vT) \quad \text{and} \\ T' = \gamma(T - vL) \end{array} \right\} \quad (6)$$

In matrix form the system of equations (5) is:

$$\begin{bmatrix} L \\ T \end{bmatrix} = \begin{bmatrix} \gamma & \gamma v \\ \gamma v & \gamma \end{bmatrix} \begin{bmatrix} L' \\ T' \end{bmatrix} \quad (7)$$

We verify that $L'^2 + (iT')^2 = L^2 + (iT)^2$. This satisfies the first cardinal principle.

A Conceptual Error in the Calculation of Length Contraction In Special Relativity

On inverting the matrix equation we receive

$$\begin{bmatrix} L' \\ T' \end{bmatrix} = \begin{bmatrix} \gamma & (-)\gamma v \\ (-)\gamma v & \gamma \end{bmatrix} \begin{bmatrix} L \\ T \end{bmatrix} \qquad (8)$$

This corresponds to the system of equations (6).

These are the correct equations. Particularly,

$L = \gamma(L' + vT')$ (correct).

$L = \gamma L'$ (wrong).

The conceptual error here consists in treating $\gamma(vT')$ erroneously as being identically equal to zero.

Consider, for example, in a given equation: $A = B + C$, the term C [here $\gamma(vT')$] is not identically zero. If we remove this term C from the equation, an inequality: $A \neq B$ will result, not another equation.

We shall, next, explore the consequences of the correct equations on length alteration. Two tables are provided in the next two pages to bring out the interesting consequences.

A Conceptual Error in the Calculation of Length Contraction In Special Relativity

TABLE for *wrong* approach

Wrong approach
If T were zero, i.e., Δt were zero, t is a fixed moment of time, definite time, then:
L' = γL (wrong)
L'/L = γ > 1, contraction from proper length L' to L, observed length.
Wrong: T is never zero for a rigid rod

L'	T	|T| > |L|	$S^2 = L^2 - T^2$		v	v^2	$\sqrt{1-v^2}$	γ	$L = \gamma L' + \gamma vT$	$T = \gamma vL' + \gamma T$	|T| > |L|	$S'^2 = L'^2 - T'^2$	$S'^2 = S^2?$	same as first wrong L' = γL	colou wrong mm?	L'/L	Observed/ proper (wrong) L/L'	wrong L'>L or L'<L'
0.10	0.20	Yes	-0.03		0.50	0.25	0.75	1.15	0.23	0.29	Yes	-0.03	Yes	0.27	No	1.15	0.87	Contraction (wrong)
0.20	0.30	Yes	-0.05		0.50	0.25	0.75	1.15	0.40	0.46	Yes	-0.05	Yes	0.47	No	1.15	0.87	Contraction (wrong)
0.10	-0.15	Yes	-0.01		0.50	0.25	0.75	1.15	0.03	-0.12	Yes	-0.01	Yes	0.03	No	1.15	0.87	Contraction (wrong)
0.20	-0.25	Yes	-0.02		0.50	0.25	0.75	1.15	0.09	-0.17	Yes	-0.02	Yes	0.10	No	1.15	0.87	Contraction (wrong)

(negative T for the *past*)

149

A Conceptual Error in the Calculation of Length Contraction In Special Relativity

TABLE for correct approach

Correct approach
For a rigid rod, T is always > 0. Particularly, $T \geq L$.
Thus,
$L' = \gamma(L - vT)$ shall be used with a non-zero T. (Correct).
$L'/L = \gamma[1 - v(T/L)]$

proper L	proper T	$\|T\| \geq \|L\|$	$S'^2 = L^2 - T^2$	v	v^2	$1-v^2$	$\sqrt{1-v^2}$	γ	observed $\gamma L + vT$	observed $T = \gamma vL + \gamma T$	$\|T\| \geq \|L\|$	$S^2 = L^2 - T^2$	$S'^2 = S^2$?	correct $L' = \gamma L - \gamma vT$	same as first mm?	correct L'/L	observed / proper (correct) L/L'; (dilation>1)	correct L'>L or L<L'
0.10	0.20	Yes	-0.03	0.50	0.25	0.75	0.87	1.15	0.23	0.29	Yes	-0.03	Yes	0.10	Yes	0.43	2.31	DILATION (correct)
0.20	0.30	Yes	-0.05	0.50	0.25	0.75	0.87	1.15	0.40	0.46	Yes	-0.05	Yes	0.20	Yes	0.49	2.02	DILATION (correct)
0.10	-0.15	Yes	-0.01	0.50	0.25	0.75	0.87	1.15	0.03	-0.12	Yes	-0.01	Yes	0.10	Yes	3.46	0.29	Contraction (correct)
0.20	-0.25	Yes	-0.02	0.50	0.25	0.75	0.87	1.15	0.09	-0.17	Yes	-0.02	Yes	0.20	Yes	2.31	0.43	Contraction (correct)

(negative T for the *past*)

A Conceptual Error in the Calculation of Length Contraction In Special Relativity

Conclusion: Errors in calculations of observed length in various historical, popular and text books

One book:

At definite time $t = 0$ *in the unprimed frame, the position of the integer* $x' = 1$ *is, with respect to the unprimed frame given by* $x = \sqrt{(1 - v^2)}$.

Error: $x' = 1$ *is a mathematical projection on the primed frame* x'-*axis and is NOT a physical entity, the rigid rod. A mathematical projection can be outside the light triangle. But a physical entity is always within the light triangle, inalienable from its associated mathematical projection in time. For the rigid rod the projection on the time axis must be* $|t'| \geq 1$. *Thus, for the observer in the unprimed frame,* $x = \sqrt{(1 - v^2)}$ *represents contraction of only a mathematical projection* $x' = 1$ *and NOT that of a physical entity.*

Another book:

Some books write: $x'_2 = \gamma(x_2 - vt)$ *and* $x'_1 = \gamma(x_1 - vt)$ *for the two ends of the moving rod, and* $t = t_1 = t_2$.

The error is obvious.

Yet another book:

Length Contraction: $L = x_2 - x_1$ *is on the x-axis of unprimed frame. Since* $t'_2 = t'_1$, *we have* $dx' = \gamma\, dx$. *As* $\gamma > 1$, $dx < dx'$. *So we have length contraction.*

Time Dilation: *Two events take place at the same spot* $x'_2 = x'_1$ *in the primed frame;* $dt = \gamma dt'$.

Error: The author has chosen the unprimed frame for length contraction but has chosen the primed frame for time dilation.

A Conceptual Error in the Calculation of Length Contraction In Special Relativity

Velocity of Light and Some Related Figures For Reference

| One nanosecond is to one second is the same as one second is to 31.7 years. |

Velocity of light, c	
300,000	km/second
300,000,000	m/second
984,251,969	feet/second
0.98	feet/nanosecond,
this is nearly 1	feet/nanosecond
300	m/microsecond
3.33	microseconds/km

Conversion Table	
3.280840	feet/meter,
1.00E+06	microseconds/second
1.00E+09	nanoseconds/ second
1.00E+12	picoseconds/ second
1.00E+15	femtoseconds/second

Sustainable Development, Entropy, Madness and the Proposed Unit of Heat: *Hiroshima Atom Bomb Equivalent* (HABE)

How I Got This Idea

Many years ago, I came across Radiative Forcing, RF, which is defined as the change in the net downward radiation (solar and infrared) at the tropopause, 10 Kilometers above the earth surface, because of greenhouse gases, resulting in global warming. RF is estimated at around 2.45 watts per square meter. The heat energy, thus, raining down upon the earth surface from the tropopause every hour is some 16,000 times the heat content of the atom bomb that was dropped on Hiroshima. We may define a unit of heat, HABE, as the heat content of the atom bomb which was dropped on Hiroshima. This unit of heat is to be used irrespective of radiation, explosion or any other effect of the atom bomb. We may, thus say, 16,000 HABEs of heat energy is raining down on the earth's surface every hour as a consequence of Radiative Forcing. Hiroshima is a powerful symbol of mass destruction in the modern times and may also be used as a meaningful unit of heat to represent the global warming. HABE, as a unit of heat will carry a visual impact that is absent in other units of heat such as the terajoue. Global warming is leading to mass extinction of species, which may, eventually, lead to the extinction of the human species as well. We may adopt as a remedial measure, capping and rolling back the bloating human civilization. This we may achieve by way of mass education, equitable distribution of wealth (education and prosperity being the best contraceptives) and total dependence on solar energy, etc.

Sustainable Development, Entropy, Madness and the Proposed Unit of Heat: Hiroshima Atom Bomb Equivalent (HABE)

IT IS SHOCKING TO REALIZE that the mankind is continuing to consume natural resources at an exponentially increasing rate, while, at the same time, piously making noises of saving the environment by "sustainable development" defined in a manner which resembles "eating a cake and having it too", a contradiction in terms. Looking at the developed nations whose debts exceed their respective national GDPs, ever widening deficit budgets and decoupling of currency from the gold standard, (and, further, with modern India and China, aiming to grow at 10 and 15% annually), all indicators of reckless consumption patterns of the earth's limited resources. This may quicken possible destruction of the earth as a habitable planet. World will do well to learn from the 5000 years of continuously living civilization and culture of India as a proven model of sustainability.

Story of the Game of Chess

The game of chess was invented in India. There is a beautiful anecdote regarding the invention of the game of chess. The inventor was a mathematician who presented his invention to his king. The king was very much fascinated and offered to reward him generously. But, to the chagrin of the king, the mathematician sought only some quantity of wheat as his reward. He wanted one grain of wheat in the first square, two grains in the second square, four in the third square, eight in the fourth square, doubling the quantity of wheat for every subsequent square on the chessboard. The angry king dismissed the mathematician from his sight and ordered the chief minister to hand over the quantity of wheat by next day morning. To his utter surprise, the chief minister informed the king next day morning that such a quantity of wheat does not exist on the planet earth. This is an example of how an exponential demand (read, conspicuous consumption) overwhelms the limited resources of the earth no matter how large.

Sustainable Development, Entropy, Madness and the Proposed Unit of Heat:
Hiroshima Atom Bomb Equivalent (HABE)

<u>Law</u>: *No matter how large, no limited natural resource of a planet shall sustain an exponentially increasing demand on it for any great length of time.*

The rate of consumption of fossil fuel, for example, has increased exponentially from 4 to 10 GTOE in last fifty years. The population of the world has increased exponentially from 2 to 7 billion in the same period of time. Apart from the increase of population by nearly 4 fold, the individual's requirements have also increased, say, nearly 10 fold in the mean while. The atmospheric CO_2 has increased exponentially from 280 ppm to 350 ppm in the same period of time. The global warming has raised the average temperature of the world by about 0.5 degrees Celsius mostly within the last 50 years. The trend is exponential. Such a temperature rise had not taken place in the last 10,000 years of the geological history. *The heat energy involved in melting of ice to raise sea level by one millimeter is 1.5×10^{14} megajoules. This is 1.8 million HABEs. If we assume that the entire 100 mm of rise of sea level in the last 50 years is from melting of ice, then this represents nearly 50 million HABEs which is nearly an atom bomb for every twenty five people on the earth.*

'Hiroshima Atom Bomb Equivalent' (HABE) *Defined*

We pump in a million tons of carbon worldwide in to the atmosphere every hour i.e., a lac tons of carbon every 6 minutes or a lac tons of carbon dioxide every two minutes. This is giving rise to the enhanced greenhouse effect causing a rain of more than 4 times the heat content of the atom bomb - dropped over Hiroshima - from the stratosphere some 10 kilometers above to the earth's surface every second. Every year we commemorate the day Hiroshima was bombed. How shall we commemorate the 4 atom bombs' worth of heat raining down every second?

This has brought about the enhanced greenhouse effect which traps some of the heat that would otherwise radiate away in to space and sends it down to the earth's surface by way of

Sustainable Development, Entropy, Madness and the Proposed Unit of Heat: Hiroshima Atom Bomb Equivalent (HABE)

'RADIATIVE FORCING' or "RF". Let us define an unit of heat energy for the purpose of this essay in terms of the heat energy yielded by the atomic bomb exploded over Hiroshima which was 82 terajoules. Let us name this quantity of energy of 82 terajoules or as one unit of **HABE, Hiroshima Atomic Bomb Equivalent** of heat energy, to be used irrespective of the aspects of nuclear radiation or explosion. The heat energy that is raining down upon the earth surface from tropopause is about four HABEs every second!

The sea level worldwide has increased by around 100 mm because of global warming. Each mm rise in the sea level from melting of ice in Antarctica and Greenland represents enough heat to be equivalent to nearly 8 million atom bombs similar to the ones dropped over Hiroshima. Spread uniformly all over the globe this will be a bomb every 10 kilometers from each other. Here we consider only the heat as if accumulated gradually without any blast or nuclear effects.

Human population has grown exponentially under the favorable conditions brought about by engineering, technology, medicine which, at the same time, led to mass extinction of other species.

'Adequate Population' *Defined*

We would like to define, for the purpose of this essay, a population of 2 billion (year 1950) as an '**Adequate population**'. This population had given us our rockets and airplanes, radio and television sets, relatively and quantum mechanics, and had carried forward the rich arts, literature, philosophies and mathematics of the earlier times.

Entropy, Development and the Environment

The law of monotonous increase of entropy in isolated thermodynamic systems implies monotonous increase of disorder in the universe. When we bring in development to a location, we bring in order in the form of engineering designs, standards, quality, etc. This throws out a corresponding amount of disorder in to the neighboring environment in keeping with the law monotonous increase of entropy. When we use bulldozers and such other equipment operating on thermodynamic cycles, or electric motors that get their power ultimately from thermal power plants including the nuclear operating on thermodynamic cycles, they are all subject to the limitation of the Carnot's cycle which is the ideal thermodynamic cycle. Carnot's cycle practically ensures that we throw at least twice as much heat in to the surrounding environment as is converted in to useful work. This translates in to throwing an additional double amount of entropy, i.e., disorder in to the surrounding environment. Thus, when we bring in development to a location, we throw at least thrice times as much entropy, i.e., disorder, in to the surrounding environment as the order associated with the development.

[Entropy and Mind

I am sorry what I am going to touch upon here may appear out of context. However, I am unable to resist myself. It appears to me that we may be able to apply the concept of thermodynamic entropy conceptually to our mind as well. The idea is this: as we progressively discipline a part of our mind, we throw a correspondingly large amount of entropy and heat to the rest of it. Look at students furiously preparing for tests. Also think of the common citizen in more developed parts of the world where he has to train himself on ever growing amounts of skills, complexity and coordination of information. This is like walking several tight ropes all at the same time. No wonder, more people are stressed the more a civilization is developed. Perhaps, the concept of Gross Domestic Happiness adopted by Bhutan is superior to GDP.]

Sustainable Development, Entropy, Madness and the Proposed Unit of Heat: Hiroshima Atom Bomb Equivalent (HABE)

Large Number of Emerging Areas of Studies and Failed Technologies

We may interpret the large number of studies emerging in various areas in the last few decades as indicative of the desperation of the mankind to catch up with the large number of woes emerging from large scale interference with the environment. Exactly what is the score card as of now? If the continuing problems of the environment and the number of failed technologies are any indication, we are not doing well, nor going to do well in the foreseeable future. Consider the list of various new areas of studies first. Last twenty years saw the following partial list of studies emerge in the area of industrial maintenance: (i) breakdown maintenance, (ii) preventive maintenance, (iii) condition based maintenance, (iv) TPM, (v) six sigma quality, (vi) ISO 9k, etc., you too may add your endless lists from the legal, economic, administrative and management studies.

Next, the failed technologies! (i) Solar PV not increasing beyond 10-12% efficiency commercially for last 30 years, (ii) no superconductors carrying bulk power at zero loss, (iii) total cycle pollution load of fuel cell is no less on environment, leading to pull back of research allocations worldwide, (iv) no new technology in the automotive sector in place of the piston and cylinder engines under use for the last 150 years (the Wankel engine did not take off commercially), (v) no fusion power that was promised decades ago, (vi) no economic solution is available for storage of electricity, (vii) no economic solution found for corrosion resistance in building metals.

Let us not be unduly optimistic of our technological puissance to fix any and all problems emerging out of very large scale anthropogenic disturbance of the natural environment. Clearly the speed of destruction far outstrips the speed of proven technological quick fixes.

Sustainable Development, Entropy, Madness and the Proposed Unit of Heat: Hiroshima Atom Bomb Equivalent (HABE)

A Recipe for Escaping the Global Warming Mess

Not easy. Organizations responsible to the world should actually quantify, certify or verify that the technical or other measures being adopted in the name of sustainable development are indeed adequate to meet global goals of sustainability, one of which is elimination of global temperature rise with in a decade. Citizens should articulate such demands to responsible organizations. (1) cap and roll back the civilization, scale down the civilization to a population of two billion, an 'adequate population' by adapting education and prosperity as the best contraceptives, (2) meanwhile, plan for executing a controlled crash of the human economy which is necessarily a small subset of the environmental economy, equitable distribution of wealth shall help the process (may be for a generation an educated population shall demand comforts at the cost of the environment, but soon afterwards, a reduced population shall reverse the effect), (3) switch over to solar and wind powers, a 100% carbon free power generation, (4) divert a vast proportion of the population to cultural (soft) activities rather than civilization building (hard) activities, (5) demand demonstrable verification and certification of sustainability for proposed technologies (for illustration, a proposed technology may be certified as 40% sustainable if it is demonstrated by acceptable methods of estimation or experimental demonstration that the technology will help reduce the contribution of greenhouse gases worldwide, emanating from the corresponding sector of application by 40% per year provided that the technology is adapted 70% world over in coming 10 years' time). By a slew of several such methods we may be able to defer an early destruction of our spaceship, the planet earth.

Story of a Madman (By Way Of Conclusion)

Once, while driving his car on a deserted road, a gentleman felt that the wheels are wobbling. Getting down from his car, he saw that all the four nuts of a front wheel are missing. While he was scratching his head, he sensed that a person was beckoning him from the other side of a barbed wire fence. The person

Sustainable Development, Entropy, Madness and the Proposed Unit of Heat: Hiroshima Atom Bomb Equivalent (HABE)

advised him to remove a nut from each of the remaining three wheels and use these three nuts on the fourth wheel and drive on to the next town and complete the repairs. The gentleman thanked the person but, somehow, was puzzled. On, confirming that the barbed wire fencing encloses a lunatic asylum, he wondered aloud, how a mad fellow could offer a workable solution for repairing the car which did not occur to him. The inmate of the lunatic asylum told him that 'I may be mad but I am not an idiot.'

This is, more or less, the story of the human civilization, today, which has gone mad and is fast moving to a collective suicide, pushing the planet way beyond its carrying capacity. This form of madness is apart from the other well-known madness related to the accumulated capacity for thermonuclear destruction of the world many times over, MAD: mutually assured destruction! Our earth is a space ship, a sphere some twelve thousand kilometers across, hurtling through space some three million kilometers a day, complete with a fragile but adequate life supporting system which we are doing everything to destroy. The environment is so poisonous today that it no more supports the life of large animals like the jungle cats, vultures, and many other life forms in substantial numbers as it did earlier. Can time be very far away for the poison and destruction to boomerang on the Homo sapiens? Let us refrain from destroying our beautiful blue-green space ship in frenzy. There no other known planet like this within a hundred light years.

Sustainable Development, Entropy, Madness and the Proposed Unit of Heat: Hiroshima Atom Bomb Equivalent (HABE)

Exponential Behavior of Environmental Parameters

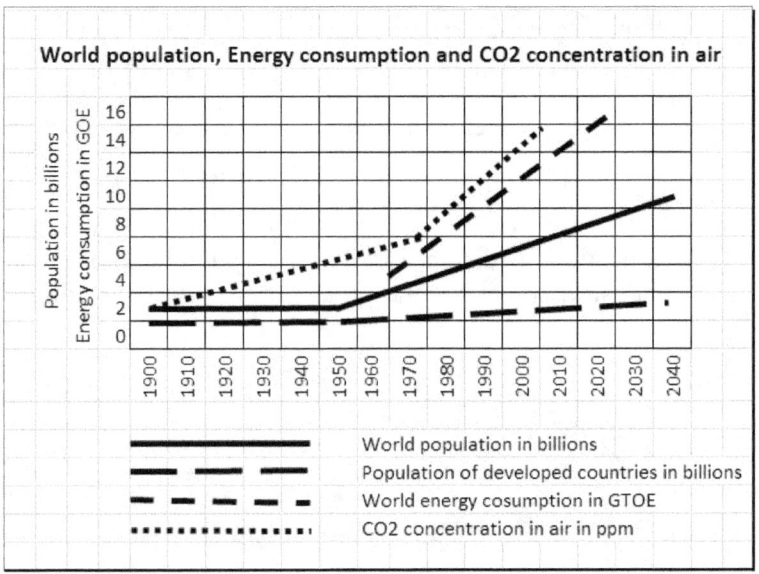

"At present, we are stealing the future, selling it in the present, and calling it GDP".

-Paul Hawken

Sustainable Development, Entropy, Madness and the Proposed Unit of Heat: Hiroshima Atom Bomb Equivalent (HABE)

The Root Cause of Environmental Disturbances

Engineering and technology
   |

Population→Deforestation   →Loss of bio diversity
   | |→Soil erosion→ Earthquakes

    |→Food, housing, clothing, transportation, governance, education, medicare, fossil fuel consumption on a vast scale for industry, power generation and transportation leading to energy crisis
   |

Air pollution
   |

Enhanced greenhouse effect
   |

Global warming
|→Sea level rise-→ Earthquakes (partly) →Tsunami
Sun strokes, super cyclones, flash floods, etc.

> *"Gross National Happiness is more important than GDP".*
>
> *- HM Jigme Singye Wangchuk, King of Bhutan*

An Accounting Framework for Presenting Specific Energy and Energy Conservation Figures for an Industrial Plant

How I Got This Idea

I have seen many discussions on energy conservation where specific consumptions based on total production are discussed passionately and fruitlessly where explanations are demanded by those in authority and lame excuses offered by those who are not. Nobody realizes that there are large varieties of specific consumptions and these are not absolute quantities but are mere indices, that some are useful for setting up a project but wholly others are useful for saving money while operating the plant. Poring over all these aspects I developed an accounting framework for presenting specific energy and energy conservation figures for energy audit of an industrial plant.

IT IS NECESSARY TO FORMULATE an acceptable accounting framework for presenting specific energy and energy conservation figures for energy audit of an industrial plant. Energy conservation for an industrial plant must primarily answer the question: how much more we'd have had to pay extra for energy had we not undertaken certain energy conservation activities. There may be a dozen alternative choices for specific energies. A proper choice is very important as a wise choice shall render unnecessary many explanations in future.

Errors of Misrepresentation

The following two types of errors are likely in energy conservation estimates based on (i) year-end-results for plants having multiple products where (ii) specific energy is based solely on the dominant product (say product A):

Error of TYPE- I

As the specific energy is based on product A by choice, an increase in the production of a lesser product (say product B) may result in a larger consumption of energy in the year-end-result and may be erroneously perceived as a loss to the plant.

Error of TYPE- II

A decrease in the production of a lesser product (say product B) may result in a lesser consumption of energy in the year-end-result and may be erroneously perceived as a saving.

Thus, choice of specific energy with respect to one or more dominant products is NOT a good choice for estimating energy conservation figure for a facility/unit. Such a choice of specific energy cannot help us to trace the gain or loss to either efficient or otherwise operation of any equipment or systems of equipment. (However such a choice of specific energy based on one or a few dominant products may be useful for examining commercial viability of selecting one out of several alternative processes before establishment of a plant).

An accounting framework shall be acceptable for presenting specific energy and energy conservation figures for an industrial plant only if it is free or substantially free from both the types of errors I & II and, further, if the framework will provide figures that can be traceable to efficient or otherwise operation of equipment or system of equipment.

Issue 1: Additional Transport Equipment to Ensure Supply of an Input Material from a Larger Distance

Sometimes it may become inevitable to ensure supply of raw materials from a larger distance. Whereas this may speak on the commercial viability of a process, this has nothing to do with the efficient operation of installed equipment. This is a necessary expenditure of energy and does NOT constitute wastage of energy. Specific energy based solely on the dominant product (say product A) shall result in Error of TYPE- I.

Recommendation
Calculate specific energy after correcting for energy spent on extra transportation equipment.

Issue 2: Intermediate Buffer Stock

The energy expended to create an intermediate buffer stock shall inflate the year-end-result and should be deducted before specific energy is estimated based on the dominant product (say product A). Otherwise the result will be an Error of TYPE- I.

Recommendation
Calculate specific energy figure for each plant separated by the intermediate buffer stock. Also, this must be declared in the disclosure of practices in the accounting framework.

Issue 3: Correct Choice of Specific Energy

For plants having multiple products there will be a large variety of specific energies possible. This is because one may choose the basis from among many possibilities like (i) the dominant product, say A; (ii) simple sum of all the products, say $A + B + C + D + $ --; (iii) weighted sum of all or some of the products, $aA + bB + cC + dD + $ -- where a, b, c, d, etc. are the chosen weights; (iii) the weights may be chosen as the market price which may vary from year-to-year; (iv) the feed-to-plant; etc. Nearly every one of these will be asked for or used from time-to-time for various purposes of comparison or explanation. However specific energy figures based on one or more products in a multi-product plant will lead to loading these figures arbitrarily with overhead energy consumption from the products which are not represented in the basis for specific consumption). Even where all products are considered, the questions of (i) proper weightage and (ii) energy used for producing rejects, which is apt to be large (considering that waste shall generally travel from the beginning to the end of the process consuming energy all the way), are unsettling issues. Still more interesting is a particular reject which suddenly finds a value in the market or a product which is suddenly losing the market.

Recommendation
Choose specific energy on FEED- TO- PLANT basis.

AN ACCOUNTING FRAMEWORK

Financial Year Under Audit: 20__ - 20__

Energy Conservation

Summary Statement: Electricity/ FO/ Coal/Gas/Water, etc. (Separate Estimates)

Plant	Choice of analysis: (Prorata or nominal)
Plant 1	Prorata
	(based on chosen specific energy)
Plant 2	Prorata
Utility 1	Prorata
Utility 2	Prorata
Housing Colony	Nominal
(based on year-to- year difference)	
Admin. Building:	Nominal
	Total

Disclosures

1) Type of analysis:
(a) Prorata (based on specific consumption; suitable for large installations with measurable output),

(b) Nominal (yr-to-yr differences for small installations) as in the Summary statement above: Electricity/ FO/ Coal/Gas/Water, etc.

2) Plant wise practices:
a) Whether figures are presented after deducting energy expended for extra transportation,

b) Whether figures are presented after deducting extra equipment installed in the meanwhile,

c) Whether estimated figures (i.e., those which are not recorded from calibrated instruments) are presented in *italics,*

d) Whether the specific energy figures are on feed-to-plant basis, or based on the dominant product/ a few of the more important products (plain sum or weighted sum; basis of weighting),

e) Policy/ vision/ strategy adopted by the unit: Annex #

f) Analysis of energy scenario of the unit in terms of 'Energy Cost/Total Cost' and 'Energy Cost/ Variable Cost': Annex #

g) Energy audits undertaken equipment wise if any: Annex #

h) Energy conservation activities undertaken in the plants: Annex #

i) Statement on instruments (existing/ working/ calibration): Annex #

j) Improvements made in instrumentation if any: Annex #

k) Further disclosures if any: Annex #

A presentation of energy conservation figures in an accounting frame work such as the above may be jointly certified by the energy conservation and the finance department as representing a fair picture of energy conservation at the unit before the same is submitted for further action elsewhere.

Each plant should be asked to evolve its own practices of presenting energy conservation figures and declare these in the space for disclosures and certification. A panel of engineers and accountants may also be considered to help evolve practices suitable to each plant. The practices are generally likely to be different for each industrial plant.

Not the Last Word

No doubt, with adoption and use, the accounting framework shall evolve over time. And save valuable hours of executive altercation.

Tyranny of Errors

My Own Suffering is the Beginning

I suffer from poor memory. Consequently I suffer in the hands of my merciless examiners, superiors, juniors and my good friends and colleagues. I attempt to hide behind the adage "It is human to err". This article is an offshoot from this deficiency of mine.

INEFFICIENCY, ERRORS, MISTAKES, blunders, mischief and crime: these words describe progressively higher degrees of human failure in performance. By way of illustration, we present the following table:

Characteristic		**Exhibited by**
Inefficiency	:	Bulk machinery; petrol and diesel cars (~40% efficiency); transformers, motors and generators (80-99% efficiency; electric incandescent lamps (5% upwards) etc.
Errors	:	Instruments; wrist watch, thermometers, energy metersflow meters, pressure gauges (0.1 to 10% errors) etc.
Mistakes	:	Human being. "It is human to err".
Blunders	:	Very conspicuous mistakes.
Mischief	:	Creativity, love, play, friendship exhibit mischief as an element.
Crime	:	Harmful acts.

It is our intention to argue that whereas crimes must be punished with exemplary severity, mischief should be tolerated with a degree of nobility of spirit, blunders must not be repeated, mistakes must be accepted as a part of human frailty, and errors and inefficiencies must be lived with as necessary evils of the material world.

When a First Class engineer is recruited by a management it is known clearly that the candidate would not have received 100% marks in the qualifying examination.

The degree certificate of the engineer is the equivalent of the name plate capacity of equipment. This declares his nominal performance. If he has 70% marks, he is expected to perform up to 70% of perfection. He is rated to fail up to 30% times. He will commit 30% errors in judgment, planning and execution. Even with experience the error shall not be zero.

A professor in literature, when writing a letter, occasionally corrects his spellings and sentences. Can you hold this against his scholarship? No. This type of mistakes is a part of the intellectual machinery, common to all human beings. You cannot deny the professor his scholarship on account of these mistakes. Just as a machine handling bulk power has a certain irreducible amount of inefficiency, just as an instrument handling information has a certain irreducible amount of error as illustrated in the above table, so has a human being, handling a job, a certain amount of irreducible amount of mistakes committed by him.

It is a myth that if everybody commits mistakes, the world will fall apart. It will not as can be seen from what follows:

In an hierarchical organization, say, X reports to Y who, in turn, reports to Z and nominally, each one is, say, 70% efficient in handling his allotted job. In an ordinary scenario, as follows, mistakes actually go on reducing as the job moves from person to person.

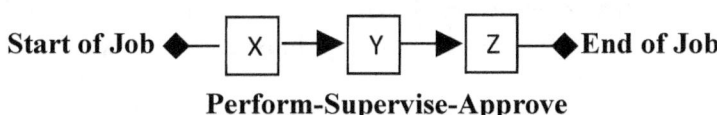

Perform-Supervise-Approve

The job is performed by X, supervised by Y and, finally, approved by Z. At each stage, assume, 30% mistakes are made by X, Y and Z respectively. A mistake will go past both X and Y if and only if X commits the mistake (30% of the times) and Y over looks the same (30% of the times). This means 30% of 30%, i.e. 9% (say, 10%) of the mistakes will go past Y's supervision. And the job will be properly executed with up to 90% correctness.

At the stage of approval, Z will commit his share of 30% mistakes. 30% of 10% is 3% ; i.e., 3% mistake will got past the approval stage. The job will be executed with 97% correctness. The world will not fall apart because of human mistakes.

In the real world, engineering processes, where proportional automatic control concept is applied, the control function operates on the basis of errors. The whole system, i.e., the plant along with the automatic control system, can function beautifully within say, about 2 to 3% error.

Mistakes will be committed by man. Punishment for mistakes will only make him hide his mistakes or make him seek to put the blame on someone else. System will lose the benefit of the consequent corrective effort. What is worse, the hidden errors will, in any case, do their invisible damage in course of time.

Sri Krishna says to Arjuna in the Gita: (BG 18:48)

"Sahajam Karma Kaunteya sadosam api na tyajet Sarbarmbha hi doshena dhumenagniribavritah".

"Every initiative has some blemish in it like fire is covered up with smoke: nonetheless, Arjuna! Perform your duty! As work, according to ones call of duty, must be undertaken".

In the language of this essay we may translate the same as follows: "Notwithstanding the possibility of committing mistakes, Arjuna! Perform, because there is no work that can be done without some mistake"

Amen.

Time Estimation of a Job

How I Got the Basic Idea

Since my college days I used to make a number of inventions (except for one, none other was of any use to any one). I found I could not meet the target date as promised to my professors despite hard work, careful estimation and follow up. I observed, later, that I used take normally three times as much time as pledged. So I look to multiplying the estimated ideal time with three before committing completion time to my professors. In my industrial career I finally understood the underlying reason.

How a Job Proceeds

A JOB NORMALLY INVOLVES several agencies none of which works at 100% efficiency. Besides, in many organizations various formalities and the prevailing work culture pre-empts a proactive effort which ultimately results in a series operation. That is to say, one agency does a part of the work and only after that another agency shall take up another part.

For the purpose of this essay let us assume that every agency has an overall efficiency of 70% . In what follows 'T' is the ideal time estimated for the job (which may tacitly but wrongly assume 100% efficiency on the part of the various agencies involved in a job).

As an example, take a job where only one agency is involved, like, for example, a motor re-winding shop. The following would happen:

$\eta(A)$ = 70%

JOB ⟶ [A] ⟶ 70% JOB in 'T': 100% JOB in 1.5 T

Time Estimation of a Job

Another example

Say, the engineering department has to complete one stage of a work using (A) departmental facilities before entrusting the balance to the agency of (B) a local contractor. Then the following would happen: (Series operation):

$\eta(A) = 70\%$ $\eta(B) = 70\%$

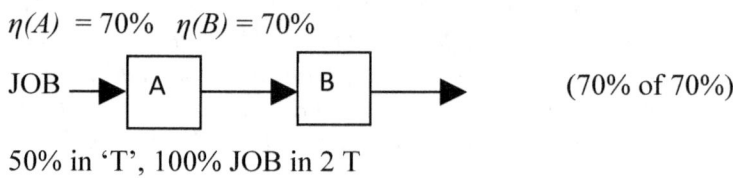

(70% of 70%)

50% in 'T', 100% JOB in 2 T

Generalizing, we would infer as follows:

AGENCIES	**% of job in T**	**100% job in**
A	70%	1.5T
A &B	70% *70% = 50%	2.0T
A,B&C	70% *70% *70% = 35%	3.0T
A,B,C & D	25%	4.0T

Real life frequently presents 3 agencies for a sizeable job:

Example A: 1. Initiator, 2. Workshop, 3. Contractor

Example B: 1. Organizer, 2. Middleman, 3. Service outlet

We can represent the above calculations in a graphical form (see the Figure below :)

Time Estimation of a Job

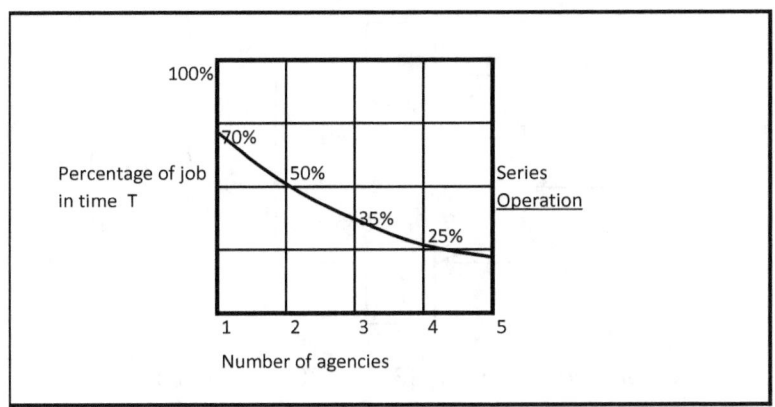

A solution is visible from the above discussion:

Parallel Operation of a Job

Say a manager can effectively call upon any or all of a number of facilities to do a particular job:

Example C: 1) Workshop
             2) Contractor

Example D: 1) Automobile section.
             2) Private taxi service.

Example E: 1) Major contractor
             2) Local contractor

In such cases the job will get done faster. This is known from experience. Let us make a model:

Time Estimation of a Job

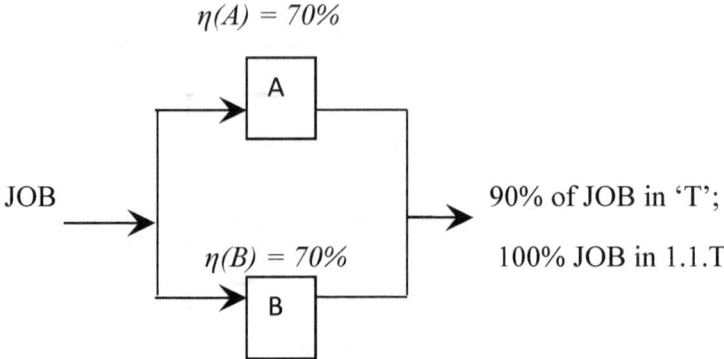

(*Calculation:* 100% - [30% x 30%] = 90%)

Say the Auto section can give a vehicle 70% of the times on demand. If on a specific demand they cannot provide a vehicle (i.e. 30% times) then there is the taxi service (also at 70% availability) to fall back upon. A vehicle will NOT be available if both the arrangements fail, i.e., 30% * 30% =10% times a vehicle will not be available. In this way a parallel arrangement will ensure availability of vehicles up to 90%. If yet another taxi service is also recruited then the NON availability will be only 10% * 30* = 3% i.e. 97% times a vehicle will be available.

If we generalize the above example we see, graphically, the following:

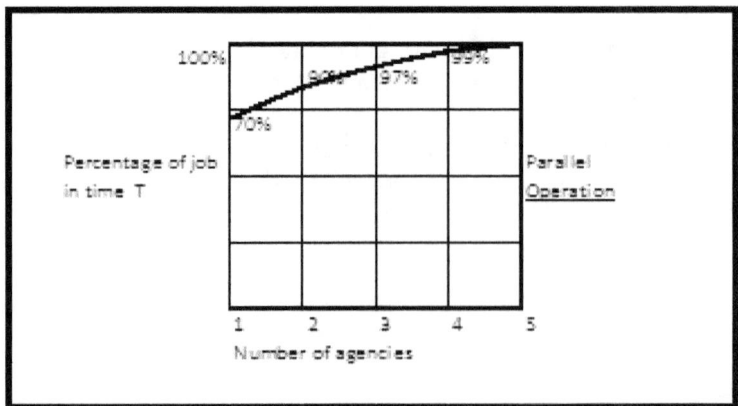

Analytically (as different from the graphical) η^N is the net percentage efficiency in *series operation* of the N number of agencies involved, and $[100 - (100-\eta)^N]$ is the net percentage efficiency in *parallel operation* of the N number of agencies involved.

The Reality behind Claims of 100% Efficiency in Job Management

Whenever there are claims of 100% efficiencies in any of the departments in an industry one may know that there are hidden parallel arrangements like, for example, rate contractor for motor rewinding shop, or, local labor contractor or a taxi service, or, the very best, a willing boss readily okaying various proposal to engage various service providers.

Industrial Organization Modeled as a Pump–Pipeline Network

How I Got This Idea
This chapter may be read along with the next one: $P = Q \times R$ and an industrial organization. The ideas of both the chapters are concurrent.

Basic Idea

SEVERAL ASPECTS OF an industrial organization can be intuitively modeled after a pump-pipe line network (hereafter called, simply, 'pumping network'). This model, like any other model, physical or intuitive, should not be stretched too far. It is important to remember that every model is a mental object and, as such, can be stretched or assaulted up to a limit beyond which it will breakdown; that is to say, every mental object, too, has an yield point (static loading) and a fatigue limit (dynamic loading) just like any material object. It is common experience that good and sound ideas are, indeed, destroyed every day around us by being stretched too far.

Work flows in an industry just like water flows in a pipe line network. An executive in his area of responsibility can be linked to a motor-pump set. An executive derives authority from the organizational charter and drives his department just as a motor drives a pump while deriving electric power from a power source.

Industrial Organization Modeled as a Pump–Pipeline Network

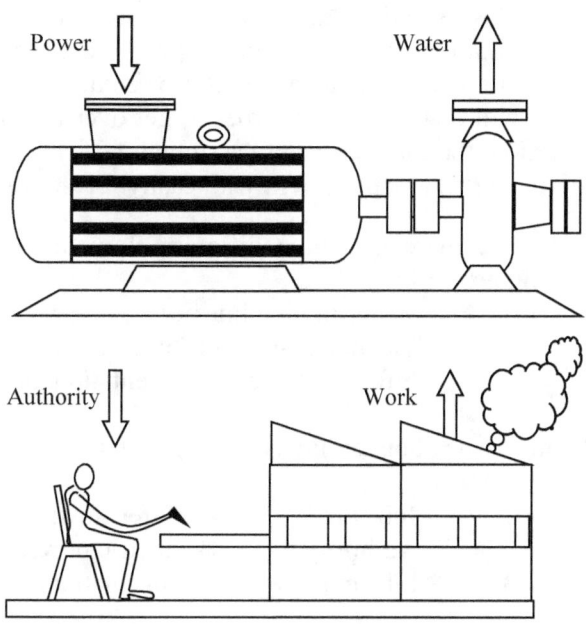

Let us ask a few questions and see if the pumping network model can come up with correct answers, which can satisfactorily explain the workings of an industrial organization. After all, that is the very purpose of toiling with a scientific model. Say, the downstream pumps do not work. The flow will, consequently, reduce. If the top pump tries to effect the designed flow everywhere, and is not adequately protected from this suicidal effort, it will generate all the pressure. This will lead to great pressure near the top pump, cause leakage, heating and mechanical damage. Same thing for a top executive in an organization where down the line executives do not or cannot (owing to distrust) exercise their authority in order to 'discharge' (mark this word) their responsibility. And just like idling pumps such executives can become drags on an organization instead of being drivers.

Industrial Organization Modeled as a Pump–Pipeline Network

In a pumping network if more and more valves are put for the purpose of monitoring and control, then, beyond a point, more obstruction will result in the path of flow leading to further pressure losses and leakages which, in turn, will lead to further power requirement. Same thing happens to the flow of job if more and more top management interference comes into play. A time will come when, in order to get over the confusion and frustration, the top man has to assume total load of supervision, engineering and management with easily foreseeable results.

We note, there are three different flows associated with a pumping network:
- (a) Water in pipe lines:
- (b) Electric power in cables;
- (c) Information in instrumentation and control

Same is true about an industrial organization:

1. Job flow ⇔ water flow;
2. Authority ⇔ electric power;
3. Information ⇔ information.

A Diversion

Consider the flow of information. What is the pressure that drives the flow of information?

It is hoped the pump-pipeline model as proposed above shall prove helpful to grasp the workings of an organization intuitively.

P = Q x R and an Industrial Organization

How I Got This Idea

I don't think I can really say how I got this idea. It was a stroke intuition. Perhaps unconsciously I looked for the pressure behind the flow of work and, then, logically, got around to various components of the net pressure of work.

Pump-Pipeline Model for an Organization

MATHEMATICALLY an organization can be modeled after a pump-pipe line network where flow of work resembles the flow of fluid. (One might, if one likes, model an organization after an electrical network with voltage sources and impedances.) Simple and useful consequences appear from such an analysis. This may help remove many of the common ills that plague the organizations and improve productivity which is, largely flow of work.

We propose the following law for organizations relating the flow of work to various pressures motivating flow of work and resistances impeding the same:

Organizations P = Q * R		(Law)
Pressure of work =	*Flow rate of jobs* *	*Resistance to the flow of jobs*

It is a very simple relationship among quantities of flow of jobs and resembles similar relationships in various branches of science and engineering. For example:

Hydraulics: $\quad\quad\quad\quad P = Q * R$
Pressure $\quad$ = Flow rate of water * Resistance in pipe line;

Electricity: $\quad\quad\quad\quad V = I * R$
Voltage $\quad$ = $\quad$ Electric current * Conductor Resistance;

Thermodynamics: $\quad\quad T = Q * R$
Temperature = $\quad$ Flow rate of heat * Thermal Resistance.

We notice that, generally, where there is a flow there is a pressure driving the flow against the resistance in the path of the flow. (As an interesting diversion, think of the train of thoughts incessantly going on in your mind, involuntarily. Discover the 'pressure' driving the 'flow of thoughts'. Obviously there is a pressure because the thinking does get accelerated to a high pitch when you are under great pressure. Besides, it takes a great deal of exertion to tame or stem the flow of thought. This may possibly start you on a path of *Raja yoga*.)

One sees that our daily language is a rich reservoir of intuitive models and these can be gainfully employed for understanding life. Art precedes, and succeeds science.

Pressure $\quad$ = $\quad$ *Flow rate* $\quad$ * $\quad$ *Resistance to the*
of work $\quad\quad\quad\quad$ *of jobs* $\quad\quad\quad\quad$ *flow of jobs:*

$\quad P = Q * R \quad\quad\quad\quad\quad\quad$ (Law)

Let us look at the components of each term of the above relationship:

<u>Pressure of work</u>:

$\quad\quad \Sigma Pi = P1 + P2 + P3 +$ etc.; where

P1 = Authority, P2 = Target, P3 = Search for excellence etc.

<u>Flow rate of 'work</u>;
$\quad\quad$ So many tons of products per shift;

So many inspections conducted per day,

So many vehicles repaired per week, etc.

In every case we see a quantity referred to a fixed period of time.

Resistance to work:
This is the most interesting part. In fact this is the very animal we are after to improve productivity in the organization. This has got several static and dynamic components. First, the static components:

$R1$ = demoralized workforce,

$R2$ = un-streamlined procedures,

$R3$ = aging of plant and machinery,

$R4$ = propensity for negative decisions,

$R5$ = executives who dodge work,

$R6$ = workers who dodge work, etc.

Next the dynamic components:

$R7$ = Opposition to change (this is similar to 'inertia of mass' or 'inductance of electric inductor')

$R8$ = hoarding instinct (this is similar to a 'reservoir' or an 'electric capacitor'; the tendency to hoard unnecessarily large amounts of materials in various sub stores in the fear of a possible shortage; this will prevent optimum availability of materials to all concerned at the right time and right place and, thus, shall hamper the flow of work).

We, further, see that the stated law yields

$$Q = P/R$$

this implies that reduction of resistance to work shall raise the output.

It shall call for great sensitivity to discern all the various components of resistances, both static and dynamic components, and predict their combined effect to trouble shoot an organization which is not doing well in terms of job flow rate.

What Next?

Set up management consultancies. Complicate an obviously simple idea. Make money.

A Structural Limitation of Descriptive Language: One Dimensionality

How I Got This Idea

I think, I suddenly realized one day, that an entire book can indeed be written down in a single long straight line. Much the same is true of the spoken language as one word follows the other in a strict sequence. This feature of one dimensionality constitutes an important geometrical structure of the descriptive languages used by the mankind. Various consequences of this limitation of one-dimensionality of the descriptive languages and various attempts to overcome the same are discussed.

Basic Role of Language

THE BASIC ROLE OF LANGUAGE is to transmit a mental image from Mr. A to Mr. B. We can model the process thus:

| Mental image of A | *Language* → | Mental image of B |

One can use the above mentioned model to identify several limitations of language inherent in its role as a vehicle of mental image.

Language is an artificial creation of the human mind. It is still developing. Every year dozens of words are created, borrowed and used with different connotations, and modifications. Our tacit assumption that language can express everything is not correct. One notices how a person falters, pauses, hesitates and then corrects himself, while giving words to an important thought or feeling for the first time.

A Structural Limitation of Language

Language introduces errors in the information it is supposed to transmit. The totality of the distortion in the transmission of signals is called NOISE by the communications engineer. All information carrying devices like the radio, the T.V., the computer are also subject to NOISE; this is also true about the language as a medium of transmission of mental imagery. This is one reason why we correct our statements, and sometimes entirely scratch and rewrite, letters, essays and poems.

Limitation of One-Dimensionality of Descriptive (Verbal and Written) Language

A further limitation of the descriptive language arises from the fact that it has a one dimensional structure. Anything that can be put on a very long straight line is inherently one dimensional. You can write a book on a thin strip of paper of perhaps ten kilometers in length. One word comes after another and is followed by exactly one more word. In contrast to the descriptive language, a map is two dimensional and an idol is three dimensional.

The one dimensionality of structure of descriptive language makes it difficult to express our experiences which are multi-dimensional. To one who has never seen an aero plane, it is difficult to describe the experience of a flight. However, aided with pictures, the job is easier. It is easiest if one gets hold of a wooden model of an aero plane. However, this is a trivial example. The really beautiful and powerful examples of successful attempts by man at surmounting this limitation of one dimensionality of descriptive language are: poetry, drama, song, dance, mathematics, humor, jokes and the like. For example when one says, "A bird in hand is worth two in bush" one is neither speaking about a bird nor a bush. If one actually tries to describe the idea to another in a plain manner, one will take full five minutes.

To give yet another beautiful example, consider Walter de la Mare's poem "Napoleon":

A Structural Limitation of Language

"What is this world O soldiers! It is I.   I,
this incessant snow, this northern sky.
Soldiers! This solitude through which we go is I."

This, if we take it purely literally, is pure nonsense. Nevertheless the poet has been able to express high spirituality by overcoming the limitation of one dimensionality of descriptive language.

Of humor I shall give no examples.

I will give two more examples to show how man has attempted successfully to overcome the limitation of one dimensionality of language. One is pure music, instrumental, which are without words. The other is what the physicist Dr. Douglas R. Hofstadter calls breaking the mind of logic in his celebrated book *GÖDEL, EISCHER, BACH: AN ENTERNAL GOLDEN BRAID*. The example is a Koan. Koan is an anecdote form used by Zen Buddhists to shock pupils into great revelations. I have lifted, with due apologies, a Koan, a commentary and a poem by Mumon a great Zen sage, and a further commentary by Dr. Hofstadter from his book.

"Koan:

A monk asked Nansen: "Is there a teaching no master ever taught before"?

    Nansen said "yes, there is".

    "What is it?" asked the monk.

Nansen replied. "It is not mind, it is not Buddha, it is not things"

    Munon's commentary:

Old Nansen gave away his treasure words. He must have been greatly upset.

A Structural Limitation of Language

Munon's Poem:

Nansen was too kind and lost his treasure. Truly, words have no power. Even though the mountain becomes the sea, words cannot open another's mind.

In this poem Munon seems to be saying something very central to Zen, and not making idiotic statements. Curiously, however, the poem is self-referential and thus it is a comment not only on Nansen's words, but also on its own ineffectiveness. This type of paradox is quite characteristic of Zen. It is an attempt to "break the mind of logic". As well as transcending the structural limitation of one dimensionality of descriptive language.

It seems better to leave off with the beautiful excerpts from Douglas Hofstadter's.

INDEX

Acceleration, 5, 22, 73, 190
Aircraft, 110ff
Ambition, 7
Anger, 7
Atman, 38
Atom bomb, 153ff
Aurobindo, 99
Awareness, 37ff, 49

Beauty, 21, 35, 41
Brain, 13ff, 28, 52, 60ff, 69, 94, 98

Carbon, 155, 159
Chanting, 18, 55, 97
Chhandogya upanishad, 89
Civilization, 11, 27, 53, 99ff, 153ff, 157ff
Concentration, 18
Concept, 88ff, 94, 100, 106, 110, 132, 157
Cyclone, 25, 57ff, 96, 162

Dance, 39, 40, 186
Death, 15, 31, 36, 63,
Dirac, 66
Dynamic, 9, 26, 55, 58, 81ff, 136, 178, 183

Ego, 5ff, 17, 29, 34, 38, 51, 58, 76ff, 79, 89ff, 99
Einstein, 65, 123, 128

Élan vital, 28
Elastic, 55ff
Electron, 25, 66, 84
Elementary particles, 88
Emotion, 35ff, 48ff, 55
Environment, 18, 26, 29, 56, 94, 99, 102, 154ff

Fatigue, 58, 178
Feedback, 87ff
Feeling, 37, 40, 51, 78, 89, 185
Fight, 6
Flow, 24, 49ff, 54, 101, 109, 113, 136, 169, 178ff
Flywheel, 55, 58, 107
Force, 5ff, 22ff, 43ff, 56, 73, 83, 103ff, 124, 141
Fracture, 88ff
Fundamental (*also see* Incidental), 21ff, 35, 36, 64, 72, 75, 107ff, 123ff

Gravitation, 10, 22ff, 73, 81, 135

Heart, 57ff
Heat, 7, 22, 50ff, 73, 153ff, 179, 182
Homo sapiens, 16ff, 160
Homoeopathy, 91ff
Honesty, 21, 35
Hydraulics, 182

Incidental (*also see* fundamental), 21, 36
Industrial, 26, 163ff, 173, 178ff
Inefficiency, 169ff,
Inertia, 9, 22ff, 55ff, 65, 73, 76, 81ff, 124, 141ff, 183
Information, 34, 41, 53, 79, 90, 157, 170, 180, 186
Insect, 13ff
Instinct, 27, 37
Intelligence, 11ff, 29, 37, 76
Intuition, 37, 60, 128, 181
Isha Upanishad, 30ff

Koan, 187

Language, 38ff, 50, 185ff
Length contraction, 132ff

Manifestation, 51, 71ff, 86ff
Mantra, 55ff, 97
Maya, 9, 49, 58, 82, 88
Meditation, 9, 18, 27, 51
Mischief, 169
Music, 40, 187
Mystical, 41

Nature, 13ff, 43, 66, 72ff, 81, 88, 94, 102, 124, 130, 135,
Newton's laws, 5ff, 73ff, 86, 106ff, 123ff,
Noise, 9, 41, 113, 154, 186

Observer, 37, 65ff, 73, 122, 133
Organization, 170ff
Osho, 31

Poem, 39ff
Population, 155ff
Potential, 7, 21ff, 34, 55, 83ff,
Prana, 28
Pressure, 49ff, 179ff

Qi, 28
Quality, 76, 79ff, 102, 148, 157ff
Quantum mechanics, 66ff, 73, 89, 156

Radiative Forcing, 153ff, 156
Relativity, 65, 73, 127ff, 142
Religion, 18, 70
Resistance, 50ff, 101, 181ff
Revenge, 6, 102

Self, 6, 15, 20, 28ff, 48,
　　51, 58ff, 70, 85ff
Soul: *see* Self

Thought, 35, 48ff, 70,
　　182, 185

Unknown , 31
Unknowable, 31

Wave, 25, 66ff, 88, 107,
　　126, 135ff
Wave-function collapse, 66
Wave-packet, 66